PHYSICS RESEARCH AND TECHNOLOGY

UNDERSTANDING TIME EVOLUTION

PHYSICS RESEARCH AND TECHNOLOGY

Additional books and e-books in this series can be found on Nova's website under the Series tab.

PHYSICS RESEARCH AND TECHNOLOGY

UNDERSTANDING TIME EVOLUTION

ASGER S. THORSEN
EDITOR

Copyright © 2020 by Nova Science Publishers, Inc.

All rights reserved. No part of this book may be reproduced, stored in a retrieval system or transmitted in any form or by any means: electronic, electrostatic, magnetic, tape, mechanical photocopying, recording or otherwise without the written permission of the Publisher.

We have partnered with Copyright Clearance Center to make it easy for you to obtain permissions to reuse content from this publication. Simply navigate to this publication's page on Nova's website and locate the "Get Permission" button below the title description. This button is linked directly to the title's permission page on copyright.com. Alternatively, you can visit copyright.com and search by title, ISBN, or ISSN.

For further questions about using the service on copyright.com, please contact:
Copyright Clearance Center
Phone: +1-(978) 750-8400   Fax: +1-(978) 750-4470   E-mail: info@copyright.com

NOTICE TO THE READER

The Publisher has taken reasonable care in the preparation of this book, but makes no expressed or implied warranty of any kind and assumes no responsibility for any errors or omissions. No liability is assumed for incidental or consequential damages in connection with or arising out of information contained in this book. The Publisher shall not be liable for any special, consequential, or exemplary damages resulting, in whole or in part, from the readers' use of, or reliance upon, this material. Any parts of this book based on government reports are so indicated and copyright is claimed for those parts to the extent applicable to compilations of such works.

Independent verification should be sought for any data, advice or recommendations contained in this book. In addition, no responsibility is assumed by the Publisher for any injury and/or damage to persons or property arising from any methods, products, instructions, ideas or otherwise contained in this publication.

This publication is designed to provide accurate and authoritative information with regard to the subject matter covered herein. It is sold with the clear understanding that the Publisher is not engaged in rendering legal or any other professional services. If legal or any other expert assistance is required, the services of a competent person should be sought. FROM A DECLARATION OF PARTICIPANTS JOINTLY ADOPTED BY A COMMITTEE OF THE AMERICAN BAR ASSOCIATION AND A COMMITTEE OF PUBLISHERS.

Additional color graphics may be available in the e-book version of this book.

Library of Congress Cataloging-in-Publication Data

ISBN: 978-1-53617-874-6

Published by Nova Science Publishers, Inc. † New York

CONTENTS

Preface		vii
Chapter 1	Semiquantum Time Evolution: Classical Limit, Dissipation and Quantum Measurement *A. M. Kowalski and A. Plastino*	1
Chapter 2	Semiquantum Time Evolution II: Density Matrices *A. M. Kowalski and A. Plastino*	25
Chapter 3	Objective and Nonobjective Mathematical Description of the Electric Charge Transport *Agneta M. Balint and Stefan Balint*	55
Chapter 4	Objective and Nonobjective Mathematical Description of the Mechanical Movement of a Material Point, Due to the Use of Different Type of Fractional Order Derivatives *Agneta M. Balint and Stefan Balint*	99
Index		131

PREFACE

Understanding Time Evolution first considers that the evolution of quantum operators is canonical with the total Hamiltonian, and that the generator of the temporal evolution of the classical variables is the mean value of this Hamiltonian, evaluated with a purely quantum Density Matrix.

The authors introduce the general MaxEnt Density Matrix for systems where quantum and classical degrees of freedom interact. This methodology can describe the interaction between microscopic and macroscopic degrees of freedom.

Next, the objectivity of the mathematical description of electric charge transport is explored. It is shown that the description of electric charge transport using fractional order derivatives is non objective.

Similarly, the closing study explores the mathematical description of mechanical movement.

As explained in Chapter 1, the interplay between quantum and classical systems, is a subject of current interest. Consider two coupled systems. If quantum effects in one of them are small with regards to the other, that one regards as classical greatly simplifies the description and yields deep insight into the composite system's dynamics. In this methodology, the authors consider that the evolution of quantum operators is canonical, with the total Hamiltonian and that the generator of the temporal evolution of the classical variables, is the mean value of this Hamiltonian, evaluated with a purely quantum Density Matrix. As a result, all the properties of the quantum system are fulfilled at any time, even when considering dissipation. This methodology can not only represent a good approximation, but it can be the one that describes the interaction between microscopic and macroscopic systems, as in the case of the interaction of a quantum system with a bath or in that of a measurement process.

In Chapter 2, the authors introduce the general MaxEnt Density Matrix for systems where quantum and classical degrees of freedom interact. Consider two coupled systems. One is quantum for sure. If quantum effects in the other one are small with regards to those of the first system, the second can be regarded as classical, which greatly simplifies the description. This methodology can not only represent a good approximation, but it can be the one that describes the interaction between microscopic and macroscopic degrees of freedom. The authors consider that the evolution of quantum operators is the canonical one and that the generator of the temporal evolution of the classical variables is the mean value of the total Hamiltonian. As a result, all the properties of the quantum system are fulfilled at any time. In particular, the Density Matrix always verifies the Liouville's equation, describing situations where the Principle of Uncertainty coexists with Chaos and Dissipation.

In Chapter 3 the objectivity of the mathematical description of the electric charge transport is discussed. It is shown that the description of the electric charge transport in electric circuit, across passive and active biological neuron membrane, along passive and active neuron axons and dendrits and in biological neuron networks, using Caputo or Riemann-Liouville fractional order derivatives defined with integral representation on finite interval, is nonobjective. It is shown also that the mathematical description of these phenomena using integer order derivatives, general Liouville-Caputo or general Riemann-Liouville fractional order derivatives defined with integral representation on infinite interval, is objective.

In Chapter 4 it is shown that, in the material point mechanics the mathematical description using Caputo or Riemann-Liouville fractional order derivative defined with integral representation on finite interval is nonobjective and, the mathematical description using general Caputo Liouville or general Riemann-Liouville fractional order derivative defined with integral representation on infinite interval, is objective.

In: Understanding Time Evolution
Editor: Asger S. Thorsen

ISBN: 978-1-53617-874-6
© 2020 Nova Science Publishers, Inc.

Chapter 1

SEMIQUANTUM TIME EVOLUTION: CLASSICAL LIMIT, DISSIPATION AND QUANTUM MEASUREMENT

A. M. Kowalski[*] *and A. Plastino*
CICPBA and Depto. de Fisica-IFLP, FCE,
Universidad Nacional de la Plata,
La Plata, Argentina

Abstract

The interplay between quantum and classical systems, is a subject of current interest. Consider two coupled systems. If quantum effects in one of them are small with regards to the other, that one regards as classical greatly simplifies the description and yields deep insight into the composite system's dynamics. In this methodology, we consider that the evolution of quantum operators is canonical, with the total Hamiltonian and that the generator of the temporal evolution of the classical variables, is the mean value of this Hamiltonian, evaluated with a purely quantum Density Matrix. As a result, all the properties of the quantum system are fulfilled at any time, even when considering dissipation. This methodology can not only represent a good approximation, but it can be the one that describes the interaction between microscopic and macroscopic systems, as in the case of the interaction of a quantum system with a bath or in that of a measurement process.

---
[*]Corresponding Author's Email: kowalski@fisica.unlp.edu.ar.

Keywords: semiclassical, semiquantum, classical limit, quantum dissipation, quantum measurement

1. INTRODUCTION

The interaction between quantum and classical degrees of freedom is indeed a subject of much current interest that describes semiquantum time evolution. As one gathers from the above statements, this interplay can be pictured with regards to two coupled systems. If quantum effects (QE) in one of them, called let us say, system 1, are small in comparison with the QE of the other, system 2, it makes sense to consider the former as classical, which greatly simplifies the description and yields deep insight into the composite system's dynamics. As an approximation, its use is historical. Examples can be found, such as Bloch equations [1], Jaynes-Cummings semi-classical model [2, 3], collective nuclear motion [4], etc.

In this vein, we need to point out a fundamental difference between us and authors that have dealt with semiclassical scenarios. We consider, as will be seen below, that there the evolution of quantum operators is the canonical and that the generator of the temporal evolution of the classical variables is the mean value of the quantum Hamiltonian H [5, 6, 7, 8, 9, 10, 11]. As a result, all the properties of the system 2 remain purely quantum at any time, even when considering dissipation (for example, the uncertainty principle (UP)).

This approach not only represents a good approximation, but it could conceivably be, as insinuated in preceding lines, the very one that describes the interaction between microscopic and macroscopic systems, as in the cases of (i) the interaction of a quantum system with a bath [8, 9, 10, 11] or (ii) a measurement process [5, 6, 12]. We will emphasize here the second aspect and show that semiquantum evolution provides an exactly solvable model for the quantum measurement problem.

Two applications will be considered. First, the classical limit of the quantum dynamics and, second, quantum dissipation, which in turn is apt for modelizing the quantum measurement problem.

Let us emphasize that the classical-quantum transition constitutes an extremely important physics topic. In such regards, quantum mechanics' classical limit (CLQM) is certainly a frontier issue [13, 14, 15, 16, 17]. In this vein, people regard "quantum" chaotic motion as deserving concentrated attention. Zurek and others [15, 16, 17] have enlightening discussed how our classical reality can emerge from the quantum world. We will contribute here, hopefully, our grain of salt to such a discussion.

The classical limit of chaotic dynamics has been analyzed in terms of a celebrated semi-quantum model in which quantum degrees of freedom interact with classical ones [18, 8]. Both system were investigated in detail A) from a purely dynamic viewpoint in [8] and also B) using statistical quantifiers derived from Information Theory (IT) in [19, 20, 21, 22].

On the other hand, the concept of quantum friction is an intriguing theme of current interest. Different attempts to quantify dissipative forces have received attention, as experimental evidence has been accumulating with regards to the presence of dissipation phenomena in several microscopic processes and in Quantum Computation [23]. In spite of the fact that several techniques have been employed [24, 25, 26], an unanimously accepted prescription for quantifying dissipating systems has not yet been devised. The main criticisms that have been made allude to the fact that they promote an apparent violation of Heisenberg's uncertainty principle.

Here, we will tackle this issue (and hopefully overcome such criticism) by slightly modifying the motion equations by the addition of a new dissipative term to them.

2. THE MODEL

We consider the interaction between a quantum system and a classical one described by a Hamiltonian of the form

$$\hat{H} = \hat{H}_q + H_{cl}\hat{I} + \hat{H}_{cl}^q, \qquad (1)$$

where $\hat{H}_q$ and H_{cl} stand for quantal and classical Hamiltonians, respectively, and $\hat{H}_{cl}^q$ is an interaction potential. $\hat{I}$ is the Identity operator. The dynamical equations for the quantum observables are the canonical ones [8], i.e., any operator $\hat{O}$ evolves in the Heisenberg picture as

$$\frac{d\hat{O}}{dt} = -i\hbar[\,\hat{H},\hat{O}\,]. \qquad (2)$$

The concomitant evolution equation for its mean value is $\langle\hat{O}\rangle \equiv \text{Tr}\,[\hat{\rho}\,\hat{O}(t)]$ is

$$\frac{d\langle\hat{O}\rangle}{dt} = -i\hbar\langle[\,\hat{H},\hat{O}\,]\rangle, \qquad (3)$$

where all mean values are taken with respect to a proper quantum density operator $\hat{\rho}$. $[\hat{H}, \hat{O}_i]$ can always be cast as

$$[\hat{H}, \hat{O}_i] = i\hbar \sum_{j=1}^{q} g_{ji} \hat{O}_j, \qquad i = 0, 1, \ldots, q, \qquad (4)$$

for $q \to \infty$. We are interested in finite q-values. In such an instance, the set $\hat{O}_i$ closes a partial Lie algebra with the Hamiltonian H. In (4), the coefficients g_{ji} (conforming a matrix G), depend don the classical variables. Consider then the evolution of two classical variables to be called A and P_A. It obeys classical Hamiltonian equations of motion, generated by the expectation value of the quantum H, $\langle \hat{H} \rangle$, i.e., [11]

$$\frac{dA}{dt} = \frac{\partial \langle \hat{H} \rangle}{\partial P_A}, \qquad (5a)$$

$$\frac{dP_A}{dt} = -\frac{\partial \langle \hat{H} \rangle}{\partial A}. \qquad (5b)$$

The complete set of equations (3) + (5) constitute an autonomous set of coupled first-order ordinary differential equations (ODE). They allow for a dynamical description in which no quantum rules are violated, i.e., the commutation-relations are trivially conserved for all times, since the quantum evolution is the canonical one for an effective time-dependent Hamiltonian (A and P_A play the role of time-dependent parameters for the quantum system) and the initial conditions are determined by a proper quantum density operator $\hat{\rho}$ [11].

Now, if in place of (5), we consider for the classical variables the set of equations

$$\frac{dA}{dt} = \frac{\partial \langle \hat{H} \rangle}{\partial P_A}, \qquad (6a)$$

$$\frac{dP_A}{dt} = -\frac{\partial \langle \hat{H} \rangle}{\partial A} - \eta P_A, \qquad (6b)$$

we face a dissipative system. The parameter $\eta > 0$ is a dissipative one, and plays a prominent role in the present considerations. Of course, the second term on the r.h.s. of (5b) appears there in the ad-hoc fashion usually employed in describing classical friction. Through this parameter η, the classical variable is coupled to an appropriate reservoir. Energy is dissipated into this reservoir. Consequently, the last term on the r.h.s. of (1) allows one to think of "quantum dissipation", albeit via an indirect route: the quantum system (for example, a degree of freedom of a system) interacts with a classical one (the rest of the system, whose behavior may be considered as classical) which, in turn, is coupled to the reservoir (the environment). The central idea of this methodology is that of discussing quantum friction using this indirect route, which allows for a dynamical description in which no quantum rules are violated.

We consider in this work a peculiar space, that will be referred to as the "u-space", in order to pursue our investigations. The set of equations derived from Eqs. (3) (for variables belonging to the quantal system) and from (6) (for the classical variables) configure an autonomous set of first-order coupled differential equations of the form

$$\frac{d\vec{u}}{dt} = \vec{F}(\vec{u}), \tag{7}$$

where $\vec{u}$ is an appropriate, generalized variable (a "vector" with both classical and quantum components). If ones considers an arbitrary volume element V_S enclosed by a surface S in the concomitant u-space, the dissipative η term induces a contraction of V_S [27] (the divergence of $\vec{F}$ is easily seen to be $-\eta$ since the matrix G in the set of equations (4) is traceless, on account of the canonical nature of Eqs. (3). One is thus led to

$$\frac{dV_S(t)}{dt} = -\eta V_S(t), \tag{8}$$

which entails that our system is a dissipative one [27]. If the classical Hamiltonian adopts the general appearance

$$H_{cl} = \frac{1}{2M} P_A^2 + V(A), \tag{9}$$

one easily ascertains that the temporal evolution for the total energy $\langle \hat{H} \rangle$ is given by

$$\frac{d\langle \hat{H} \rangle}{dt} = -\frac{\eta}{M} P_A^2, \tag{10}$$

whose significance is to be appreciated in the light of Eq. (8). The commutation-relations are again trivially conserved for all time (the quantal evolution is the canonical one), so that one is able to avoid any quantum pitfall.

3. MATTER–FIELD INTERACTION AND CLASSICAL LIMIT

We are concerned below with a bipartite system representing the zero-th mode contribution of a strong external field to the production of charged meson pairs [18, 8]. The associated Hamiltonian reads

$$\hat{H} = \frac{1}{2}\left(\frac{\hat{p}^2}{m_q} + \frac{P_A^2}{m_{cl}} + m_q \omega^2 \hat{x}^2\right), \tag{11}$$

with the quantity

$$\omega^2 = \omega_q^2 + e^2 A^2, \tag{12}$$

responsible for the introduction of *non linear terms* that originate chaos [11]. $\hat{x}$ and $\hat{p}$ above are quantum operators, while A and P_A are classical canonical conjugate variables. The term ω^2 is an interaction one introducing nonlinearity, with ω_q a frequency. m_q and m_{cl} are masses, corresponding to the quantum and classical systems, respectively.

As shown in [8, 22], using Eqs. (3) and (5), one has to deal with the autonomous system of nonlinear coupled equations:

$$\frac{d\langle \hat{x}^2 \rangle}{dt} = \frac{\langle \hat{L} \rangle}{m_q}, \tag{13a}$$

$$\frac{d\langle \hat{p}^2 \rangle}{dt} = -m_q \omega^2 \langle \hat{L} \rangle, \tag{13b}$$

$$\frac{d\langle \hat{L} \rangle}{dt} = 2\left(\frac{\langle \hat{p}^2 \rangle}{m_q} - m_q \omega^2 \langle \hat{x}^2 \rangle\right), \tag{13c}$$

where the quantum-classical interaction takes place via (12). Using (5) we obtain for A and P_A the relations

$$\frac{dA}{dt} = \frac{P_A}{m_{cl}}, \tag{14a}$$

$$\frac{dP_A}{dt} = -e^2 m_q A \langle \hat{x}^2 \rangle. \tag{14b}$$

Eqs. (13) and (14) constitute an autonomous system of non linear coupled equations.

So as to investigate the classical limit one needs also to consider that classical counterpart of the Hamiltonian (11) in which all variables are classical

$$H_{class} = \frac{1}{2}\left(\frac{p^2}{m_q} + \frac{P_A^2}{m_{cl}} + m_q\omega^2 x^2\right). \quad (15)$$

In such case, Hamilton's equations lead to a classical version of (13). For the classical variables, i.e., the subset $\{x^2, p^2, L = 2xp\}$, and the macroscopic variables $\{A, P_A\}$, we have

$$\frac{dx^2}{dt} = \frac{L}{m_q}, \quad (16a)$$

$$\frac{dp^2}{dt} = -m_q\omega^2 L, \quad (16b)$$

$$\frac{dL}{dt} = 2\left(\frac{p^2}{m_q} - m_q\omega^2 x^2\right), \quad (16c)$$

$$\frac{dA}{dt} = \frac{P_A}{m_{cl}}, \quad (16d)$$

$$\frac{dP_A}{dt} = -e^2 m_q A x^2. \quad (16e)$$

A very important quantity for our endeavors, E_r, will be introduced next. The classical limit is obtained considering the limit of a "relative energy" [8]

$$E_r = \frac{|E|}{I^{1/2}\omega_q} \to \infty, \quad (17)$$

where $E = \langle \hat{H} \rangle$ is the total energy of the quantum system and I is an invariant of the motion described by the system (13). I relates to the Uncertainty Principle:

$$I = \langle \hat{x}^2 \rangle \langle \hat{p}^2 \rangle - \frac{\langle \hat{L} \rangle^2}{4} \geq \frac{\hbar^2}{4}, \quad (18)$$

and describes the deviation of the semiquantum system from the classical one given by $I = 0$. Note that Eqs. (13) do not explicitly depend on $\hbar$. The mean values depend on it only implicitly, via I, through the initial conditions. To tackle this system one needs to appeal to a numerical treatment. The pertinent analysis is effected by plotting quantities of interest for E_r, that ranges in $[1, \infty]$.

A few words are in order here in connection with the meaning of our all important quantifier E_r, essentially the ratio between the total energy of the quantum system and an invariant of the motion I, related to the uncertainty principle. Thus, E_r measures a sort of quantum energy/uncertainty ratio. The smaller the ratio, the more classical the composite system becomes. This is a rather original view of classicity. It intensifies whenever the uncertainty diminishes in comparison to the available energy. Paradoxically, if it diminishes to a sufficient degree, chaos ensues. Contrarily, whenever energy becomes contaminated by uncertainty to an enough degree, chaos dies. Thus, chaos becomes here viewed as a surplus of uncontaminated (by uncertainty) energy.

3.1. Numerical Results

In our numerical study we used $m_q = m_{cl} = \omega_q = e = 1$. For the initial conditions needed to tackle Eqs. (13), we employed $E = 0.6$. Thus, we fixed E and then varied I in order to determine the distinct values for E_r. We employed 55 different values for I. Further, we set $\langle L \rangle(0) = L(0) = 0$, $A(0) = 0$ (for the quantum and for the classical instances), while $x^2(0)$. $\langle x^2 \rangle(0)$ takes values in the intervals $(0, 2E), (E - \sqrt{E^2 - I}, E + \sqrt{E^2 - I})$, with $I \leq E^2$, respectively.

From careful study (analytic and numerical) of Eqs. (13) and (14), one can obtain a considerable amount of information, some of which is reported below. For $E_r = 1$ the quantum system acquires all the available energy $E = I^{1/2}\omega_q$ and the quantal and classical variables get located at the fixed point ($\langle \hat{x}^2 \rangle = I^{1/2}/m_q\omega_q$, $\langle \hat{p}^2 \rangle = I^{1/2}m_q\omega_q$, $\langle \hat{L} \rangle = 0$, $A = 0$, $P_A = 0$) [8]. Since $A = 0$, the two systems become uncoupled. For $E_r \sim 1$ the system is of an almost quantal nature, with a quasi-periodic dynamics [8].

As E_r augments, quantum features are rapidly lost and a semiclassical region is entered. From a given value E_r^{cl}, the morphology of the solutions to Eqs. (13) begins to resemble that of classical curves [8]. One indeed achieves convergence of Eqs. (13)'s solutions to the classical ones.

We regard as semiclassical the region $1 < E_r < E_r^{cl}$. Within such an interval we highlight the important value $E_r = E_r^{\mathcal{P}}$, at which chaos emerges [20]. Finally, for very large E_r–values, the system becomes classical. Both types of solutions (classical and semiquantal) coincide

Important results are depicted in Figure 1, where we depict some relevant Poincaré surfaces (sections' cuts with $A = 0$) corresponding to the system of equations conformed by (13) and (14). In the graph we plot $\langle \hat{p}^2 \rangle$ vs. $\langle \hat{x}^2 \rangle$. Numerical details are given in the pertinent captions. The existence of chaos was verified by the calculation of the Lyapunov characteristic exponent, which

Semiquantum Time Evolution

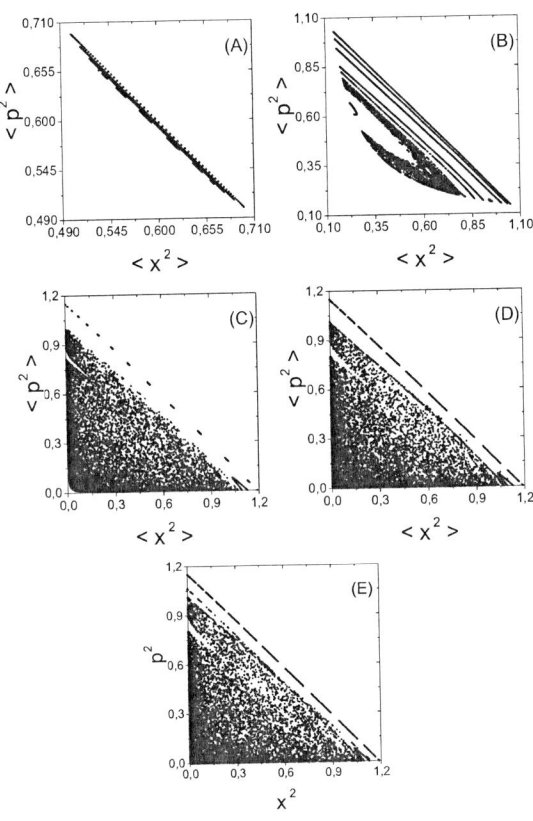

Figure 1. $A = 0$ Poincaré surfaces, calculated for $\langle \hat{p}^2 \rangle$ vs. $\langle \hat{x}^2 \rangle$. Sub-figure (A) corresponds to $E_r \simeq 1$ (quasi-quantal case), displaying periodicity and quasi-periodicity. Sub-figure (B) corresponds to the quantum-classical transition zone, where chaos emerges. Sub-figure (C) corresponds to E_r^{cl}, value of E_r at which the classical zone begins. Sub-figures (D) correspond to still larger values of E_r, where classic features become more and more prominent. Figure F) is associated to the classical instance $I = 0$. Notice the coexistence of the Uncertainty Principle with chaos. We have checked the accuracy of all results by verifying the constancy in time of the dynamical invariants E and I (within a precision of 10^{-10}).

is positive for the curves in question. We have also checked the accuracy of all results by verifying the constancy in time of the dynamical invariants E and I (within a precision of 10^{-10}). Sub-figure (A) corresponds to $E_r \simeq 1$ (quasi-quantal case), displaying periodicity and quasi-periodicity. Sub-figure (B) corresponds to the quantum-classical transition zone, where chaos emerges. Sub-figure (C) corresponds to E_r^{cl}, value of E_r at which the classical zone begins. Sub-figures (D) correspond to still larger values of E_r, where classic features become more and more prominent. We can observe the convergence to Figure F), corresponding to the classical instance $I = 0$, i.e. to the solution of the purely classical system of equations. We also observe that the Sections in Figure 1 a–e) are bounded by the curves

$$\frac{\langle \hat{p}^2 \rangle}{m_q} + m_q \omega_q^2 \langle \hat{x}^2 \rangle = 2E, \tag{19a}$$

$$\langle \hat{x}^2 \rangle \langle \hat{p}^2 \rangle = I. \tag{19b}$$

Eq. (19a), 'the "energy curve", (with $A = 0$, and $P_A = 0$) represents a periodic, stable solution for the system of equations (13). Eq. (19b) can be called the "Uncertainty Principle orbit" The classical orbits are enclosed within the region circumscribed by the classical counterpart of the curves (19) (the classical case $I = 0$ may be regarded as a "degenerate" hyperbole, that overlaps the coordinate axis). **Notice the coexistence of the Uncertainty Principle with chaos**, specially in Figure 1B).

Another way of representing things will be now described. We can always associate to our physical problem *a time-series* given by the E_r-evolution of appropriate expectation values of the dynamical variables. Given this series, one may use entropic measures and other information-quantifiers so as to characterize the process and compare such description with that given by a purely classical dynamic solution [19, 20, 22].

3.2. Using Information Quantifiers

Information theory measures and probability spaces Ω are intimately linked, since in order to evaluate statistical quantifiers one needs per force to determine the probability distribution P associated to the dynamical system or the time series under study. In our work we employ, for instance, entropies and the so called statistical complexities. An example is that of the Boltzmann-Gibbs-Shannon entropy,

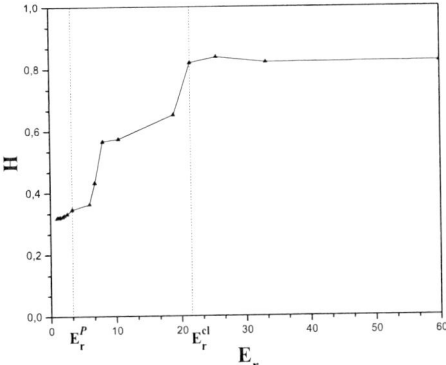

Figure 2. Entropy vs E_r. See the three concomitant regions: semiquantal, classical-quantum transition one, and classical.

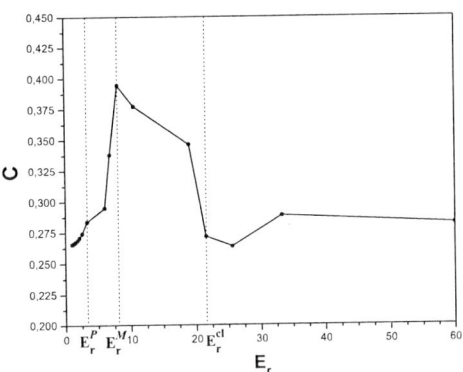

Figure 3. Statistical Complexity vs E_r. We can observe the three zones: semiquantum, transitional, and classical. Also one can appreciate the value $E = E_r^M$, where C becomes maximal.

$$S = -\sum_{i=1}^{N} p_i \ln(p_i). \qquad (20)$$

where p_i are the probabilities associated with the associated N different system-configurations. The statistical complexity is a very important and rather recent tool, that has originated a lot of work during the last 25 years ([28] and references therein). One of its main ingredients is a distance between probability distributions (PDs) (distance in probability-space) called *disequilibrium* [19]. As any distance-quantifier, it appears in many different guises. Below we will see it as the distance between the uniform probability P_e and the PD of current interest. Remark that the entropy associated to P_e is the maximum possible entropy. The standard statistical complexity measure C is then the product of the disequilibrium Q times an entropy (or a relative entropy). C adopts below the appearance [19]

$$C[P] = Q[P, P_e] \cdot H_S[P], \qquad (21)$$

where, to the probability distribution P, we associate the entropic measure $H_S[P] = S[P]/S_{max}$, with $S_{max} = S[P_e]$ ($0 \leq H_S \leq 1$). P_e is the uniform distribution and S is Shannon's entropy. The disequilibrium Q is defined in terms of a special distance in probability space called the extensive Jensen-Shannon divergence [19], being given by

$$Q[P, P_e] = Q_0\{S[(P + P_e)/2] - S[P]/2 - S[P_e]/2\}. \qquad (22)$$

with Q_0 a normalization constant ($0 \leq Q_J \leq 1$). Thus, the disequilibrium Q is here an intensive quantity (this is not true in many applications). For evaluating the probability distribution P associated to the time series (dynamical system) under study, a crucial step, we follow the methodology proposed by Bandt and Pompe (BP) (see [29] and references therein) and consider partitions of the pertinent D-dimensional space. It is well known that the BP procedure is able to "reveal" relevant details of the ordinal-structure of one-dimensional time series [29].

Figures 2 and 3 depict, respectively, the entropy H and the intensive statistical complexity C versus E_r. Our three regions are clearly appreciated. At a special value of E_r, called E_r^M, C becomes maximal. This E_r-value divides the transition zone into two sub-regions of different chaoticity.

4. DISSIPATION: A TWO-LEVEL MODEL COUPLED TO A CLASSICAL ELECTROMAGNETIC FIELD

Two-level systems coupled to a field's single-mode in a cavity constitute an important problem in several scientific fields like quantum optics. Consider

$$\hat{H} = E_1 \hat{a}_1^\dagger \hat{a}_1 + E_2 \hat{a}_2^\dagger \hat{a}_2 + \frac{\omega}{2}(p^2 + s^2) + \gamma s(\epsilon \hat{a}_1^\dagger \hat{a}_2 + \epsilon^\dagger \hat{a}_2^\dagger \hat{a}_1), \qquad (23)$$

with $E_2 > E_1$. s y p are classical variables. ϵ is chosen as a dimensionless parameter. We get a partial Lie algebra if we appropriately select relevant operators to face this problem. For instance, those pertaining to the set $\{\hat{O}_1 = \hat{a}_1^\dagger \hat{a}_1,$ $\hat{O}_2 = \hat{a}_2^\dagger \hat{a}_2, \hat{O}_3 = i(\epsilon \hat{a}_1^\dagger \hat{a}_2 - \epsilon^\dagger \hat{a}_2^\dagger \hat{a}_1), \hat{O}_4 = (\epsilon \hat{a}_1^\dagger \hat{a}_2 + \epsilon^\dagger \hat{a}_2^\dagger \hat{a}_1)\}$. Via (3) we find

$$\frac{d\langle \hat{O}_1 \rangle}{dt} = -\gamma s \langle \hat{O}_3 \rangle, \tag{24a}$$

$$\frac{d\langle \hat{O}_2 \rangle}{dt} = \gamma s \langle \hat{O}_3 \rangle, \tag{24b}$$

$$\frac{d\langle \hat{O}_3 \rangle}{dt} = -2|\epsilon|^2 \gamma s (\langle \hat{O}_2 \rangle - \langle \hat{O}_1 \rangle) + \omega_0 \langle \hat{O}_4 \rangle, \tag{24c}$$

$$\frac{d\langle \hat{O}_4 \rangle}{dt} = -\omega_0 \langle \hat{O}_3 \rangle, \tag{24d}$$

where $\omega_0 = (E_2 - E_1)$. Using Eqs. (6) we can give the classical variables the expression

$$\frac{ds}{dt} = \omega p, \tag{25a}$$

$$\frac{dp}{dt} = -(\omega s + \gamma \langle \hat{O}_4 \rangle + \eta p). \tag{25b}$$

Eqs. (24) - (25) constitute a non linear system of coupled differential equations. Remind that both classical and quantum variables are dimensionless. We take $|\epsilon| = 1$ without loss of generality. We introduce next the population-difference operator $\Delta \hat{N}$

$$\Delta \hat{N} = \hat{O}_2 - \hat{O}_1, \tag{26}$$

whose mean value measures the difference between the levels' populations ΔN. Adding to the picture the dimensionless parameter

$$\alpha = \frac{2\gamma}{\omega_0}, \tag{27}$$

we encounter the Bloch-like equations

$$\frac{d\Delta N}{d\tau} = \alpha s \langle \hat{O}_3 \rangle, \tag{28a}$$

$$\frac{d\langle \hat{O}_3 \rangle}{d\tau} = -\alpha s \Delta N + \langle \hat{O}_4 \rangle, \tag{28b}$$

$$\frac{d\langle \hat{O}_4 \rangle}{d\tau} = -\langle \hat{O}_3 \rangle, \tag{28c}$$

where $\tau = \omega_0 t$. Further addition of dimensionless parameters

$$\Omega = \frac{\omega}{\omega_0}, \qquad \delta = \frac{\eta}{\omega_0}, \tag{29}$$

allows for recasting Eqs. (25) as

$$\frac{ds}{d\tau} = \Omega p, \tag{30a}$$

$$\frac{dp}{d\tau} = -(\Omega s + \frac{1}{2}\alpha \langle \hat{O}_4 \rangle + \delta p). \tag{30b}$$

4.1. Fixed Points (FP)

We classify FP as being of type (A) or (B) according to whether the equilibrium value for s is non-null or vanishes, respectively. Appeal now to the motion invariant I_B

$$I_B = \Delta N^2 + \langle \hat{O}_3 \rangle^2 + \langle \hat{O}_4 \rangle^2, \tag{31}$$

called "Bloch-length". We find for the **type (A)** FP

$$\Delta N_f = -2\frac{\Omega}{\alpha^2}, \tag{32a}$$

$$\langle \hat{O}_3 \rangle_f = 0, \tag{32b}$$

$$\langle \hat{O}_4 \rangle_f = \pm (I_B - 4\frac{\Omega^2}{\alpha^4})^{1/2}, \tag{32c}$$

$$s_f = -\frac{\alpha}{2\Omega} \langle \hat{O}_4 \rangle_f, \tag{32d}$$

$$p_f = 0, \tag{32e}$$

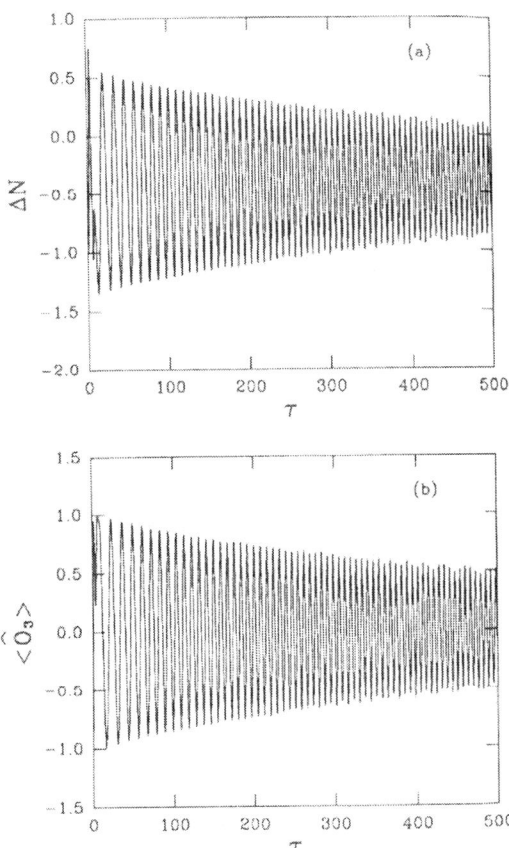

Figure 4. Time evolution of i) ΔN (a) and ii) $\langle \hat{O}_3 \rangle$ (b). We take $\delta = 1$ and $\Omega = 0.8$. The associate fixed point is $(-0.4, 0, 1.6852, -2.1065, 0)$.

that arise for $2\Omega/\alpha^2 < I_B^{1/2}$. Instead, for **Type (B)** FP we have

$$\Delta N_f = \pm I_B^{1/2}, \tag{33a}$$
$$\langle \hat{O}_3 \rangle_f = 0, \tag{33b}$$
$$\langle \hat{O}_4 \rangle_f = 0, \tag{33c}$$
$$s_f = 0, \tag{33d}$$
$$p_f = 0, \tag{33e}$$

without range-limitations. Note that the FP depend on the initial conditions only via the invariant I_B. The FP stability is determined, as usual, by linearizing (28) - (30) around the FP.

4.2. Attractors

It is important to confirm the attractor-character of our stable FPs. This character is associated to dissipative effects. In the pertinent studies one needs to vary both parameters and initial conditions. A synthesis of our results follows. We find

(A)

$$\Delta N_f = -2\frac{\Omega}{\alpha^2}, \qquad (34a)$$

$$\langle \hat{O}_4 \rangle_f = \pm (I_B - 4\frac{\Omega^2}{\alpha^4})^{1/2}, \qquad (34b)$$

$$s_f = -\frac{\alpha}{2\Omega}\langle \hat{O}_4 \rangle_f, \qquad (34c)$$

$$p_f = \langle \hat{O}_3 \rangle_f = 0, \qquad (34d)$$

within the range $0 < 2\Omega/\alpha^2 < I_B^{1/2}$ (strong coupling), and

(B)

$$\Delta N_f = -I_B^{1/2}, \qquad (35a)$$

$$s_f = \langle \hat{O}_4 \rangle_f = 0, \qquad (35b)$$

$$p_f = \langle \hat{O}_3 \rangle_f = 0, \qquad (35c)$$

within the range $2\Omega/\alpha^2 \geq I_B^{1/2}$.

Note that the "final" particle-flux runs from the upper to the lower level. The attractors depend a) on initial conditions through I_B, and b) upon α and Ω. There is an abrupt change in (34a) for

$$\frac{2\Omega}{\alpha^2} = I_B^{1/2}. \qquad (36)$$

Semiquantum Time Evolution

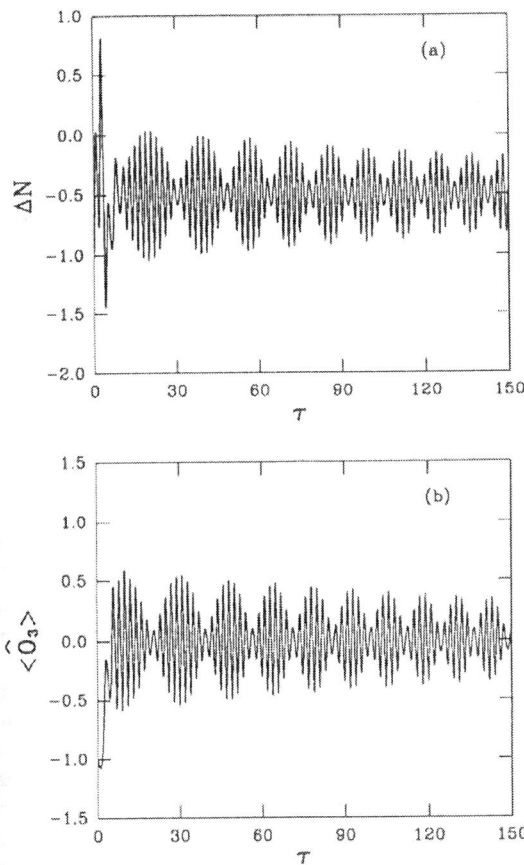

Figure 5. Time evolution of i) ΔN (a) and ii) $\langle \hat{O}_3 \rangle$ (b). We take $\alpha = 2$, $\delta = 1$, and $\Omega = 1$. The associated FP is $(-0.5, 0, 1.6583, -1.6583, 0)$.

Remark that δ does not intervene in determining the nature of the FP. This is, instead, determined by the interaction between the two systems. Attractors emerge because $\delta \neq 0$, but their location is nor correlated with their numerical values. The populations of levels $\langle \hat{O}_2 \rangle$ and $\langle \hat{O}_1 \rangle$ can be written as

$$\langle \hat{O}_2 \rangle(\tau) = \frac{1}{2}(N + \Delta N(\tau)), \tag{37a}$$

$$\langle \hat{O}_1 \rangle(\tau) = \frac{1}{2}(N - \Delta N(\tau)), \tag{37b}$$

where N is the mean value of $\hat{N} = \hat{O}_1 + \hat{O}_2$, a motion invariant. Since $\Delta N_f \leq 0$, it follows that

$$\langle \hat{O}_2 \rangle_f \leq \langle \hat{O}_1 \rangle_f, \tag{38}$$

independently of initial conditions or parameters' values. Nevertheless,

$$\Delta \langle \hat{O}_2 \rangle = \langle \hat{O}_2 \rangle_f - \langle \hat{O}_2 \rangle(0), \tag{39}$$

might be positive for type (A) if

$$\frac{2\Omega}{\alpha^2} < -\Delta N(0), \tag{40}$$

with $\Delta N(0) < 0$. This specific case describes laser excitation: the necessary energy for the transition to the upper level is provided by the classical system, represented by a single mode of frequency Ω of the field described by the conjugate variables s and p. Part of this energy is dissipated. For type (B), $\Delta \langle \hat{O}_2 \rangle \leq 0$ for any initial condition or parameter-value. Thus, s_f and the interaction potential vanish for $\tau \to \infty$. The two system uncouple at large times. Remember that in quantum scattering theory one also considers, always, the final results as taking place at a time going to ∞, which implies, in practice, minutes while microscopic times are of the order of, say, 10^{-10} seconds. Consider the expression for the variation of the quantum energy ΔE_q. One has $E_q = \langle \hat{H}_q \rangle$ and

$$\hat{H}_q = E_1 \hat{O}_1 + E_2 \hat{O}_2. \tag{41}$$

Such expression may be cast in terms of ΔN (cf. (37)) as

$$\begin{aligned}\Delta E_q &= \langle \hat{H}_q \rangle_f - \langle \hat{H}_q \rangle(0) \\ &= -\frac{1}{2}\omega_0(I_B^{1/2} + \Delta N(0)) \leq 0,\end{aligned} \tag{42}$$

where $\langle \hat{H}_q \rangle_f$ is the minimum value of $\langle \hat{H}_q \rangle$ (and also of $\langle \hat{H} \rangle$) for the given initial conditions represented by the invariant (31).

In case (A), $s_f \neq 0$. Accordingly, the interaction potential does not vanish for $\tau \to \infty$. We face the "final" Hamiltonian

$$\hat{H}_f = \frac{1}{2}\omega_0(\Delta\hat{N} - \frac{\alpha^2}{\Omega}\langle\hat{O}_4\rangle_f \hat{O}_4), \qquad (43)$$

where we do not take into account the terms (1) $(\omega_0\alpha^2/8\Omega)\langle\hat{O}_4\rangle_f^2$ and (2) $\omega_0\hat{N}/2$, which commute with our relvant operators. The above Hamiltonian describes the coupled system for $\tau \gg 1$. We have

$$\begin{aligned}\Delta E_q &= \langle\hat{H}_f\rangle - \langle\hat{H}_q\rangle(0) \\ &= -\frac{1}{2}\omega_0(\frac{\alpha^2}{2\Omega}I_B + \Delta N(0)) \leq 0.\end{aligned} \qquad (44)$$

Note that the sign of ΔE_q does nor depend upon the constant terms (1) and (2) above.

The time evolution of the expectation values of $\Delta\hat{N}$ and $\hat{O}_3$ is depicted by Figures 4 and 5, corresponding to FP of type (A) and by Figure 6, associated to FP of type (B), for $\alpha = 2$, $\delta = 1$, and $\Omega = 0.8$, 1, and 4, respectively. As stated above, the dissipative behavior intensifies for FP of type (B). As we took $\hbar = 1$, all quantities become dimensionless. We deal with a bosonic case for $N \geq 3^{1/2}$. The initial conditions are $\Delta N = -1$, $\langle\hat{O}_3\rangle = -1$, $\langle\hat{O}_4\rangle = 1$, $s = 0$, and $p = -10$.

5. THE MEASUREMENT PROBLEM

In our present considerations we found two kinds of stable fixed points of types ((A) y (B)). Those of type (A) minimize the total energy $\langle\hat{H}\rangle$ (for adequate initial conditions related to the values of the motion invariants). FP of type (B) simultaneously minimize the quantum energy $\langle\hat{H}\rangle$ and $\langle\hat{H}_q\rangle$.

The latter represents a) maximum relaxation with $\langle\hat{O}_3\rangle = \langle\hat{O}_4\rangle = s = p_s = 0$ together with maximal particle number for the lower level compatible with the initial conditions. The quantum and classical systems interact, and decouple for $\tau \to \infty$. Thus, the classical variables' values (CVV) allow one to deduce those of the quantum ones. The CCV act thus like measurement instruments. Thus, by looking at the value of the classical variables one can ascertain the value of the quantum ones. Some quantum values are correlated to the classical ones

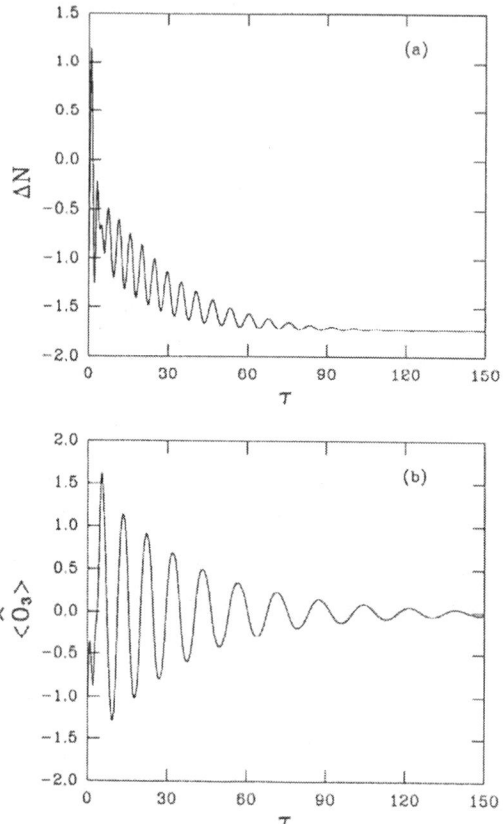

Figure 6. Time evolution of i) ΔN (a) and ii) $\langle \hat{O}_3 \rangle$ (b). We took $\alpha = 2$, $\delta = 1$ and $\Omega = 4$. The pertinent fixed point is $(-1.7321, 0, 0, 0, 0)$.

and one needs looking only at the later. as it happens in real life. We have thus achieved an exact mimic of the measurement problem.

For the type (A) of fixed points, the quantum and classical systems become entangled and cannot be separated at $\tau \to \infty$. Even if by looking at the value of the classical variables one can also learn the value of same quantum variables. On the other hand, these quantum variables do not acquire "most relaxed" values.

5.1. An Example of Measurement's Procedure

The measurement apparatus is here represented by our classical variables. The interact with the quantum ones. The experimenter waits until equilibrium is reached. This equilibrium is attained when $p = 0$ and s becomes constant. The measured value $s = s_f$ yields the following information.

(1) If $s_f \neq 0$, then

$$\Delta N_f = -2\frac{\Omega}{\alpha^2}, \quad \langle \hat{O}_4 \rangle_f = -\frac{2\Omega}{\alpha} s_f, \quad \langle \hat{O}_3 \rangle_f = 0. \tag{45}$$

Since one knows the initial number of particles N (a constant of the motion), one is also in possession of additional knowledge, namely the level populations $\langle \hat{O}_{2f} \rangle = \frac{1}{2}(N - 2\frac{\Omega}{\alpha^2})$ and $\langle \hat{O}_{1f} \rangle = \frac{1}{2}(N + 2\frac{\Omega}{\alpha^2})$.

(2) If $s_f = 0$, then

$$\Delta N_f = -I_B, \quad \langle \hat{O}_4 \rangle_f = \langle \hat{O}_3 \rangle_f = 0. \tag{46}$$

In such scenarion, knowing the value $N(0)$ and also the initial values for $\langle \hat{O}_3 \rangle$ and $\langle \hat{O}_4 \rangle$, we assess ΔN_f, $\langle \hat{O}_{2f} \rangle = \frac{1}{2}(N - I_B)$, $\langle \hat{O}_{1f} \rangle = \frac{1}{2}(N + I_B)$ plus the equilibrium quantum energy value (EQEV)

$$E_{qf} = \frac{1}{2}(E_2 + E_1)N + \frac{1}{2}(E_2 - E_1)I_B. \tag{47}$$

In case (1) the EQEV is not available, for the simple reason that the quantum and classical values remain coupled. There is no time-evolution because the composite system is located at a fixed point.

Summing up, we have presented here a schematic exact realization of the quantum measurement problem in terms of stable fixed point (attractors) dynamics.

CONCLUSION

We can summarize our main present results in the following way.

- We have shown that our semiquantum treatment does not violate any quantum rule. The time evolution is the canonical one, both for quantum operators and for classical variables. For the latter, it is generated by the mean value of the quantum Hamiltonian, taken with respect to a proper quantum density operator.

- We have enlarged the model of the preceding item so as to encompass quantum dissipation. No quantum property is violated,

- We presented a new viewpoint concerning the classical limit of our quantum processes. The evolution from a quantum region toward the classical limit is described by the percentage-amount of uncertainty in the system's energy. Classicity is reached when the percentage tends to zero.

  We have studied a system representing the zero-th mode contribution of a field to the production of charged meson pairs. The route towards classicity displays three zones: 1) semiquantum, 2) transitional and 3) classical, determined by dynamic tools and Information Quantifiers. We can also observe the convergence to the solution of the purely classical system of equations.

- We have also analyzed a two-level system coupled to a field's single-mode in a cavity. Quantum dissipation is exactly described without any quantum pitfall.

- The last fact allows us to suitably modelize the celebrated quantum measurement problem.

Acknowledgments

A. M. K. fully acknowledges support from the Comisión de Investigaciones Científicas de la Provincia de Buenos Aires (CICPBA). A. P. acknowledges support from the National research Council of Argentina (CONICET).

References

[1] Bloch, E. (1946). Nuclear Induction. *Phys. Rev.* 70, 460.

[2] Milonni, P., Shih, M., Ackerhalt, J.R. (1987). *Chaos in Laser-Matter Interactions* (World Scientific Publishing Co.: Singapore).

[3] Meystre, P., Sargent, M. (1991). *Elements of Quantum Optics* (Springer, NY).

[4] Ring, P., Schuck, P. (1980). *The Nuclear Many-Body Problem* (Springer-Verlag: Berlin, Germany).

[5] Kowalski, A.M., Plastino, A and Proto, A.N. (1995). Semiclassical model for quantum dissipation. *Phys. Rev. E* 52, 165.

[6] Kowalski, A.M., Plastino, A and Proto, A.N. (1997). A semiclassical statistical model for quantum dissipation. *Physica A* 236, 429.

[7] Kowalski, A.M., Martin, M.T., Nuñez, J., Plastino, A and Proto, A.N. (1998). Quantitative indicator for semi-quantum chaos. *Phys. Rev. A* 58, 2596.

[8] Kowalski, A.M., Plastino, A and Proto, A.N. (2002). Classical limits. *Phys. Lett. A* 297, 162.

[9] Kowalski, A.M., R. Rossignoli, R. (2018). Nonlinear dynamics of a semiquantum Hamiltonian in the vicinity of quantum unstable regimes. *Chaos, Solitons and Fractals* 109, 140.

[10] Kowalski, A.M., Plastino, A., Rossignoli, R. (2019). Complexity of a matter-field Hamiltonian in the vicinity of a quantum instability. *Physica A* 513, 767.

[11] Kowalski, A.M., Plastino, A. (2019). A nonlinear matter-field Hamiltonian analyzed with Renyi and Tsallis statistics. *Physica A* 535, 122387.

[12] Kowalski, A.M. (2016). Betting on dynamics. *Physica A* 458, 106.

[13] Halliwell, J.J., Yearsley, J.M. (2009). Arrival times, complex potentials, and decoherent histories. *Phys. Rev. A* 79, 062101:1.

[14] Everitt, M.J., Munro, W.J, Spiller, T.P. (2009). Quantum-classical crossover of a field mode. *Phys. Rev. A* 79, 032328:1.

[15] Zeh, H.D. (1999). Why Bohms quantum theory?. *Found. Phys. Lett.* 12, 197.

[16] Zurek, W.H. (1981). Pointer basis of quantum apparatus: Into what mixture does the wave packet collapse?. *Phys. Rev. D* 24, 1516.

[17] Zurek, W.H. (2003). Decoherence, einselection, and the quantum origins of the classical. *Rev. Mod. Phys* 75, 715.

[18] Cooper, F., Dawson, J., Habib, S., Ryne, R.D. (1998). Chaos in time-dependent variational approximations to quantum dynamics. *Phys. Rev. E* 57, 1489.

[19] Kowalski, A.M., Martín, M.T., Plastino, A. Rosso, O.A., (2007). Bandt-Pompe approach to the classical-quantum transition. *Phys. D* 233, 21.

[20] Kowalski, A.M., Plastino, A. (2009). Bandt-Pompe-Tsallis quantifier and quantum-classical transition. *Physica A* 388, 4061.

[21] Kowalski, A.M., Martín, M.T., Plastino, A. and Judge, G. (2014). Kullback-Leibler approach to chaotic time series. *Transactions on Theoretical Physics* 1, 40.

[22] Kowalski, A.M., Martín, M.T., Plastino, A. (2015). Generalized relative entropies in the classical limit. *Physica A* 422, 167.

[23] Weiss U. (2008). *Quantum Dissipative Systems* (Series in Modern Condensed Matter Physics) (Singapore: World Scientific).

[24] Kanai, E. (1948). On the Quantization of the Dissipative Systems. *Prog. Theor. Phys.* 3, 440.

[25] Hasse, R.W. (1978). Microscopic derivation of quantum fluctuations in nuclear reactions. *J. of Phys. A* 11, 1245.

[26] Öttinger, H.C. (2011). The geometry and thermodynamics of dissipative quantum systems Hans Christian. *EPL* 94, 10006.

[27] Arnold, V. I. (1978). *Mathematical methods of classical mechanics*. Springer, Heidelberg-New York.

[28] López-Ruiz, R., Mancini, H.L., Calbet, X. (1995). A statistical measure of complexity. *Phys Lett. A* 209, 321.

[29] Bandt, C., and Pompe, B. (2002). Permutation Entropy: A Natural Complexity Measure for Time Series. *Phys. Rev. Lett.* **88**, 174102-1.

In: Understanding Time Evolution
Editor: Asger S. Thorsen

ISBN: 978-1-53617-874-6
© 2020 Nova Science Publishers, Inc.

Chapter 2

SEMIQUANTUM TIME EVOLUTION II: DENSITY MATRICES

A. M. Kowalski[*] *and A. Plastino*
CICPBA and Depto. de Fisica-IFLP, FCE,
Universidad Nacional de la Plata, La Plata, Argentina

Abstract

In this chapter, we introduce the general MaxEnt Density Matrix for systems where quantum and classical degrees of freedom interact. Consider two coupled systems. One is quantum for sure. If quantum effects in the other one are small with regards to those of the first system, the second can be regarded as classical, which greatly simplifies the description. This methodology can not only represent a good approximation, but it can be the one that describes the interaction between microscopic and macroscopic degrees of freedom. We consider that the evolution of quantum operators is the canonical one and that the generator of the temporal evolution of the classical variables is the mean value of the total Hamiltonian. As a result, all the properties of the quantum system are fulfilled at any time. In particular, the Density Matrix always verifies the Liouville's equation, describing situations where the Principle of Uncertainty coexists with Chaos and Dissipation.

Keywords: density matrix, semiclassical, semiquantum, classical limit, quantum dissipation

---

[*]Corresponding Author's Email: kowalski@fisica.unlp.edu.ar.

1. INTRODUCTION

The interaction between quantum and classical degrees of freedom is indeed a subject of much current interest that describes semiquantum time evolution. As one gathers from the above statements, this interplay can be pictured with regards to two coupled systems. If quantum effects (QE) in one of them are small in comparison with the QE of the other, it makes sense to consider the former as classical, which greatly simplifies the description and yields deep insight into the composite system's dynamics. As an approximation, this has been used for decades. Examples can be easily found, such as Bloch equations [1], Jaynes-Cummings semi-classical model [2, 3], collective nuclear motion [4], etc.

We introduced this subject in Chapter *Semiquantum time evolution: Classical limit. Dissipation and quantum measurement* of this book *Understanding Time Evolution*. We considered there operators' and mean values' evolutions. We pointed out that a fundamental difference exists between us and other authors that have dealt with semiclassical scenarios. The diffences resides in the fact i) that we regard the evolution of quantum operators as canonical and ii) that the generator of the temporal evolution of the classical variables is the mean value of the quantum Hamiltonian H [5, 6, 7, 8, 9, 10, 11]. As a result, all the properties of the quantal system remain purely quantum at any time.

This approach not only represents a good approximation, but it could conceivably be, as insinuated in preceding lines, the very one that describes the interaction between microscopic and macroscopic systems, as in the cases of (i) the interaction of a quantum system with a bath [8, 9, 10, 11] or (ii) a measurement process [5, 6, 12]. We emphasize in the above mentioned chapter that semiquantum evolution provides an exactly solvable model for the quantum measurement problem.

Here, we introduce the general MaxEnt Density Matrix for this kind of systems. Our Density Matrix verifies the Liouville's equation for all time. Two applications will be considered. First, the classical limit of the quantum dynamics and, second, quantum dissipation.

Let us emphasize that the classical-quantum transition constitutes an extremely important physics topic. In such regards, quantum mechanics' classical limit (CLQM) is certainly a frontier issue [13, 14, 15, 16, 17]. In this vein, people regard "quantum" chaotic motion as deserving concentrated attention. Zurek and others [15, 16, 17] have enlightening discussed how our classical reality can emerge from the quantum world. We will contribute here, hopefully, our grain of salt to such a discussion.

The classical limit of chaotic dynamics has been analyzed in terms of a celebrated semi-quantum model in which quantum degrees of freedom interact with classical ones [18, 8]. Both system were investigated in detail A) from a purely dynamic viewpoint in [8] and also B) using statistical quantifiers derived from Information Theory (IT) in [19, 20, 21, 22].

On the other hand, the concept of quantum friction is an intriguing theme of current interest. Different attempts to quantify dissipative forces have received attention, as experimental evidence has been accumulating with regards to the presence of dissipation phenomena in several microscopic processes and in Quantum Computation [23]. In spite of the fact that several techniques have been employed [24, 25, 26], an unanimously accepted prescription for quantifying dissipating systems has not yet been devised. The main criticisms that have been made allude to the fact that they promote an apparent violation of Heisenberg's uncertainty principle. Here, we will tackle this issue (and hopefully overcome such criticism) by slightly modifying the motion equations by the addition of a new dissipative term to them. First of all, we review the main features of our methodology.

2. THE SEMIQUANTUM MODEL

We consider the interaction between a quantum system and a classical one described by a Hamiltonian of the form

$$\hat{H} = \hat{H}_q + H_{cl}\hat{I} + \hat{H}_{cl}^q, \tag{1}$$

where $\hat{H}_q$ and H_{cl} stand for quantal and classical Hamiltonians, respectively, and $\hat{H}_{cl}^q$ is an interaction potential. $\hat{I}$ is the Identity operator. The dynamical equations for the quantum observables are the canonical ones [8], i.e., any operator $\hat{O}$ evolves in the Heisenberg picture as

$$\frac{d\hat{O}}{dt} = -i\hbar[\hat{H},\hat{O}]. \tag{2}$$

The concomitant evolution equation for its mean value is $\langle\hat{O}\rangle \equiv \text{Tr}[\hat{\rho}\hat{O}(t)]$ is

$$\frac{d\langle\hat{O}\rangle}{dt} = -i\hbar\langle[\hat{H},\hat{O}]\rangle, \tag{3}$$

where all mean values are taken with respect to a proper quantum density operator $\hat{\rho}$. $[\hat{H}, \hat{O}_i]$ can always be cast as

$$[\hat{H}, \hat{O}_i] = i\hbar \sum_{j=1}^{q} g_{ji} \hat{O}_j , \qquad i = 0, 1, \ldots, q, \tag{4}$$

for $q \to \infty$. We are interested in finite q-values. In such an instance, the set $\hat{O}_i$ closes a partial Lie algebra with the Hamiltonian H [27, 28, 29]. In (4), the coefficients g_{ji} (conforming a matrix G), depend on the classical variables. Consider then the evolution of two classical variables to be called A and P_A. It obeys classical Hamiltonian equations of motion, generated by the expectation value of the quantum H, $\langle\hat{H}\rangle$, i.e., [11]

$$\frac{dA}{dt} = \frac{\partial \langle \hat{H} \rangle}{\partial P_A}, \tag{5a}$$

$$\frac{dP_A}{dt} = -\frac{\partial \langle \hat{H} \rangle}{\partial A}. \tag{5b}$$

The complete set of equations (3) + (5) constitute an autonomous set of coupled first-order ordinary differential equations (ODE). They allow for a dynamical description in which no quantum rules are violated, i.e., the commutation-relations are trivially conserved for all times, since the quantum evolution is the canonical one for an effective time-dependent Hamiltonian (A and P_A play the role of time-dependent parameters for the quantum system) and the initial conditions are determined by a proper quantum density operator $\hat{\rho}$ [11]. Now, if in place of (5), we consider for the classical variables the set of equations

$$\frac{dA}{dt} = \frac{\partial \langle \hat{H} \rangle}{\partial P_A}, \tag{6a}$$

$$\frac{dP_A}{dt} = -\frac{\partial \langle \hat{H} \rangle}{\partial A} - \eta P_A, \tag{6b}$$

we face a dissipative system. The parameter $\eta > 0$ is a dissipative one, and plays a prominent role in the present considerations. Of course, the second term on the r.h.s. of (5b) appears there in the ad-hoc fashion usually employed in describing classical friction. Through this parameter η, the classical variable is coupled to an appropriate reservoir. Energy is dissipated into this reservoir.

Consequently, the last term on the r.h.s. of (1) allows one to think of "quantum dissipation", albeit via an indirect route: the quantum system (for example, a degree of freedom of a system) interacts with a classical one (the rest of the system, whose behavior may be considered as classical) which, in turn, is coupled to the reservoir (the environment). The central idea of this methodology is that of discussing quantum friction using this indirect route, which allows for a dynamical description in which no quantum rules are violated.

We consider in this work a peculiar space, that will be referred to as the "u-space", in order to pursue our investigations. The set of equations derived from Eqs. (3) (for variables belonging to the quantal system) and from (6) (for the classical variables) configure an autonomous set of first-order coupled differential equations of the form

$$\frac{d\vec{u}}{dt} = \vec{F}(\vec{u}), \tag{7}$$

where $\vec{u}$ is an appropriate, generalized variable (a "vector" with both classical and quantum components). If ones considers an arbitrary volume element V_S enclosed by a surface S in the concomitant u-space, the dissipative η term induces a contraction of V_S [30] (the divergence of $\vec{F}$ is easily seen to be $-\eta$ since the matrix G in the set of equations (4) is traceless, on account of the canonical nature of Eqs. (3). One is thus led to

$$\frac{dV_S(t)}{dt} = -\eta V_S(t), \tag{8}$$

which entails that our system is a dissipative one [30]. If the classical Hamiltonian adopts the general appearance

$$H_{cl} = \frac{1}{2M} P_A^2 + V(A), \tag{9}$$

one easily ascertains that the temporal evolution for the total energy $\langle \hat{H} \rangle$ is given by

$$\frac{d\langle \hat{H} \rangle}{dt} = -\frac{\eta}{M} P_A^2, \tag{10}$$

whose significance is to be appreciated in the light of Eq. (8). The commutation-relations are again trivially conserved for all time (the quantal evolution is the canonical one), so that one is able to avoid any quantum pitfall.

In the forthcoming section, we will review the basic concepts of MaxEnt Methology. Later, we will see how to adapt a MaxEnt Density Matrix ρ to the description of our semiquantal system and then study ρ-behavior in two examples: the classical limit of the quantum dynamics and the quantum dissipation problem.

3. BASIC PROPERTIES OF A MAXENT DENSITY MATRIX

In the MaxEnt procedure [27], if we know the expectation values (EV) $\langle \hat{O}_i \rangle$ of q operators $\hat{O}_i$, the statistical operator $\hat{\rho}$ is given by

$$\hat{\rho} = \exp\left(-\lambda_0 \hat{I} - \sum_{i=1}^{q} \lambda_i \hat{O}_i \right), \tag{11}$$

where the $q+1$ Lagrange multipliers λ_i are determined so as to fulfill the set of constraints posed by our prior information (i.e., normalization of $\hat{\rho}$ and the supposedly a priori known q EV's)

$$\langle \hat{O}_i \rangle = \mathrm{Tr}\,[\,\hat{\rho}\,\hat{O}_i\,], \qquad i = 0, 1, \ldots, q, \tag{12}$$

($\hat{O}_0 = \hat{I}$ is the identity operator). A simplified way to obtain the values of the multipliers is that of solving the coupled set of equations [27]

$$\frac{\partial \lambda_0}{\partial \lambda_i} = -\langle \hat{O}_i \rangle, \qquad i = 1, 2, \ldots, q, \tag{13}$$

where

$$\lambda_0 = \mathrm{Tr}\left[\hat{\rho}\exp\left(-\sum_{i=1}^{q} \lambda_i \hat{O}_i\right)\right]. \tag{14}$$

Of course, the entropy is to be expressed, in terms of $\hat{\rho}$, according to von Neumann's definition (Boltzmann's constant is set equal to unity)

$$S(\hat{\rho}) = -\mathrm{Tr}\,[\,\hat{\rho}\,\ln\hat{\rho}\,] = \lambda_0 + \sum_{i=1}^{q} \lambda_i \langle \hat{O}_i \rangle, \tag{15}$$

and is maximized by the MaxEnt statistical operator $\hat{\rho}$ [27]. On the other hand, $\hat{\rho}$ evolves in time according to the Lioville-von Neumann equation [27]

$$i\hbar \frac{d\hat{\rho}}{dt} = [\,\hat{H}(t), \hat{\rho}(t)\,], \qquad (16)$$

If the operators entering Eq. (11) close a partial Lie algebra with respect to the Hamiltonian $\hat{H}$, i.e., (4), the Lagrange multipliers λ_j must verify the set of differential equations

$$\frac{d\lambda_i}{dt} = \sum_{j=1}^{q} g_{ij} \lambda_j, \qquad i = 0, 1, \ldots, q, \qquad (17)$$

to guarantees that the MaxEnt *form* (11) of $\hat{\rho}$ remains the same for all t [27, 28, 29]. The dynamical problem reduces itself to solving the system of equations (17), with multipliers' initial conditions calculated via Eqs. (13).

Additionally, once $\hat{\rho}(t))$ is obtained, we can determine (in the Schrödinger picture), **the temporal evolution of the EV of any operator $\hat{O}$ through**

$$\langle \hat{O} \rangle(t) = \text{Tr}[\hat{\rho}(t)\hat{O}]. \qquad (18)$$

4. MAXENT DENSITY MATRIX FOR THE SEMIQUANTUM PROBLEM

We assume total knowledge about the initial conditions of the classical variables, so that $\hat{\rho}(0)$ becomes a purely **quantum** ρ. We also assume incomplete knowledge regarding the system's quantum component. We only know the initial values of the quantum expectation values (EV) of the operators in Eq. (11).

It is important to note that in the semiquantum problem, the g_{ij} of Eq. (4) depend on the classical variables A and P_A. Having at hand $\hat{\rho}(0)$, in this case a MaxEnt density matrix, we use equation (18) (with $\hat{O} = \hat{H}$) in order to obtain $\langle \hat{H} \rangle$ and thus describe, via Eqs. (5), the temporal evolution of A and P_A. The idea is then to view the set of equations (17), together with equations (5) or (6), as constituting a single autonomous first-order system of the form $d\vec{u}_\lambda/dt = \vec{F}(\vec{u}_\lambda)$, with $\vec{u}_\lambda$ a "vector" with both classical components and Lagrange multipliers. However, the temporal evolution of the classical variables depends on the mean values $\langle \hat{O} \rangle$. This dependence can be faced using equation

(18), although introducing into our problem the nonlinearity in the multipliers that this equation implies. This nonlinearity is apparently difficult to solve. However, we will see in the two examples considered that this term is easy to handle. Thus, a pseudo phase space is thus to be considered and trajectories in this "space" ("u_λ-space"), which are the solutions of

$$\frac{d\lambda_i}{dt} = \sum_{j=1}^{q} g_{ij}\lambda_j, \qquad i = 0, 1, \ldots, q, \tag{19a}$$

$$\frac{dA}{dt} = \frac{\partial \langle \hat{H} \rangle}{\partial P_A}, \tag{19b}$$

$$\frac{dP_A}{dt} = -\frac{\partial \langle \hat{H} \rangle}{\partial A} - \eta P_A, \tag{19c}$$

$$\langle \hat{O}_i \rangle(t) = \text{Tr}[\hat{\rho}(t)\hat{O}_i], \qquad i = 1, 2, \ldots, q. \tag{19d}$$

are studied. The **no fricton case** corresponds to $\eta = 0$ in Eq. (19c). Let's go to the examples.

5. CLASSICAL LIMIT: HAMILTONIAN AND PREVIOUS RESULTS

We consider a system representing the zero-th mode contribution of a strong external field to the production of charged meson pairs [18], whose Hamiltonian is

$$\hat{H} = \frac{1}{2}\left(\frac{\hat{p}^2}{m_q} + \frac{P_A{}^2}{m_{cl}} + m_q \omega^2 \hat{x}^2\right). \tag{20}$$

where $\hat{x}$ and $\hat{p}$ are quantum operators, while A and P_A are classical canonical conjugate variables. The term $\omega^2 = \omega_q{}^2 + e^2 A^2$ is an interaction one introducing nonlinearity, with ω_q a frequency. m_q and m_{cl} are quantum and classical masses, respectively.

As shown in [7, 8], using Eqs. (3) for the set ($\hat{x}^2, \hat{p}^2, \hat{L} = \hat{x}\hat{p} + \hat{p}\hat{x}$) and Eqs. (5), one has an autonomous system of nonlinear coupled equations. We deal with such set, because it is the smallest one that carries information regarding the uncertainty principle [7, 8]. To investigate the classical limit one needs also to consider the classical counterpart of the Hamiltonian (36), in which all variables are classical. In such case, Hamilton's equations lead to a classical version of the previous semiclassical one ($\hat{L}$ is replaced by $L = 2xp$). The classical limit is obtained considering the limit of a "relative energy" [8]

$$E_r = \frac{|E|}{I^{1/2}\omega_q} \to \infty, \qquad (21)$$

where E is the total energy of the system and I is an invariant of the motion for the above mentioned system. I relates to the Uncertainty Principle:

$$I = \langle \hat{x}^2 \rangle \langle \hat{p}^2 \rangle - \frac{\langle \hat{L} \rangle^2}{4} \geq \frac{\hbar^2}{4}, \qquad (22)$$

and describes the deviation of the semiquantum system from the classical one given by $I = 0$. Note that Eqs. (2) and (3) do not explicitly depend on $\hbar$. Instead, the pertinent expectation values (EV) implicitly depend on this constant, through the initial conditions, that are here represented by I.

The pertinent dynamical analysis was performed in [8] for several quantities of interest, that were plotted against E_r, that ranged in $[1, \infty]$ [8]. In these studies, we found that our semiquantum system exhibits three regions, namely, a quasi-quantum, a classic, and a semiclassical transition zone [20]. Finally, the quantum variables were seen to converge to their purely classical counterparts. All these results have been statistically reconfirmed by using information quantifiers [19, 20, 21, 22].

6. DENSITY MATRIX AND CLASSICAL LIMIT

The set $(\hat{x}^2, \hat{p}^2, \hat{L})$ that we use is the smallest one that, satisfying Eq. (4), carries information regarding the uncertainty principle (via I). Using Eq. (13), one can determine the "initial" $\hat{\rho}(t = 0)$, whose *form* will not change in time. One has

$$\hat{\rho}(t) = \exp - \left(\lambda_0 \hat{I} + \lambda_1 \hat{x}^2 + \lambda_2 \hat{p}^2 + \lambda_3 \hat{L} \right). \qquad (23)$$

Eqs. (17) acquire now the appearance

$$\frac{d\lambda_1}{dt} = 2m_q \omega^2 \lambda_3, \qquad (24a)$$

$$\frac{d\lambda_2}{dt} = -\frac{2}{m_q} \lambda_3, \qquad (24b)$$

$$\frac{d\lambda_3}{dt} = -\frac{1}{m_q} \lambda_1 + m_q \omega^2 \lambda_2, \qquad (24c)$$

The system (24) depends in nonlinear fashion upon the classical variable A, via ω^2, which forces consideration of the corresponding classical evolution equations (5). The presence of the term $\langle \hat{x}^2 \rangle$ in the equation for P_A introduces an additional non-linearity (as such a term is a function of the multipliers) through

$$\langle \hat{x}^2 \rangle(t) = \text{Tr}[\hat{\rho}(t)\hat{x}^2], \tag{25}$$

but we will presently see that this non-linearity can be easily dealt with.

The quantity I_λ defined as

$$I_\lambda = \left(\lambda_1 \lambda_2 - \lambda_3^2\right)^{1/2}, \tag{26}$$

is, by virtue of (24), a constant of the motion, i.e., $dI_\lambda/dt = 0$. This invariant is the equivalent of the one in Eq. (22), expressed in terms of the λ's.

We need to calculate λ_0, to relate the initial values of the multipliers, and their respective EV's, using Eq. (13). We begin by performing a change of representation, made by recourse to the unitary transformation

$$\hat{X} = \frac{\sqrt{2}}{2}\left(\frac{\lambda_2}{\lambda_1}\right)^{1/4}\left(\left(\frac{\lambda_T}{\lambda_V}\right)^{1/4}\hat{x} + \left(\frac{\lambda_V}{\lambda_T}\right)^{1/4}\hat{p}\right), \tag{27a}$$

$$\hat{P} = \frac{\sqrt{2}}{2}\left(\frac{\lambda_1}{\lambda_2}\right)^{1/4}\left(-\left(\frac{\lambda_T}{\lambda_V}\right)^{1/4}\hat{x} + \left(\frac{\lambda_V}{\lambda_T}\right)^{1/4}\hat{p}\right), \tag{27b}$$

where

$$\lambda_V = \sqrt{\lambda_1 \lambda_2} + \lambda_3, \tag{28a}$$
$$\lambda_T = \sqrt{\lambda_1 \lambda_2} - \lambda_3. \tag{28b}$$

For reasons of convergence, λ_1, λ_2, and $\lambda_1\lambda_2 - \lambda_3^2$ must be positive. Then, λ_V and λ_T are positive too and I_λ in (26) is well defined. Of course, the transformation (27) preserves commutation relations. Thus, I is also preserved. The new operators $\hat{X}$ and $\hat{P}$, via the λ's that appear as coefficients in their definition, are explicitly time-dependent and contain all the relevant information regarding the classical degrees of freedom. Now $\rho(t)$ becomes

$$\hat{\rho}(t) = \exp(-\lambda_0)\exp\left[-\lambda_I\left(\hat{X}^2 + \hat{P}^2\right)\right]. \tag{29}$$

Despite the characteristics assigned to $\hat{X}$ and $\hat{P}$, the operator $\hat{X}^2 + \hat{P}^2$ has a discrete spectrum, one resembling that of a the Harmonic Oscillator, because the commutation relations are preserved for all time. After a little algebra, it is easy to see from (14) that

$$\lambda_0 = -\ln\left[\exp(\hbar\,I_\lambda) - \exp(-\hbar\,I_\lambda)\right], \tag{30}$$

and using Eq. (13) (or Eq. 18), the particular EV's can be cast in the fashion

$$\langle \hat{x}^2 \rangle = \frac{T(I_\lambda)}{I_\lambda}\lambda_2, \tag{31a}$$

$$\langle \hat{p}^2 \rangle = \frac{T(I_\lambda)}{I_\lambda}\lambda_1, \tag{31b}$$

$$\langle \hat{L} \rangle = -2\frac{T(I_\lambda)}{I_\lambda}\lambda_3, \tag{31c}$$

with $T(I_\lambda)$ given by

$$T(I_\lambda) = \frac{\hbar}{2}\left(\frac{\exp(2\,\hbar\,I_\lambda) + 1}{\exp(2\,\hbar\,I_\lambda) - 1}\right). \tag{32}$$

Further, we deduce from (31) that

$$T(I_\lambda) = \sqrt{I}, \tag{33}$$

Now we are in a position to recast our dynamical system of equations in "u_λ-space", by recourse to (31a), i.e., as a closed system of equations in both multipliers and classical variables. We have

$$\frac{d\lambda_1}{dt} = 2m_q\omega^2\lambda_3, \tag{34a}$$

$$\frac{d\lambda_2}{dt} = -\frac{2}{m_q}\lambda_3, \tag{34b}$$

$$\frac{d\lambda_3}{dt} = -\frac{1}{m_q}\lambda_1 + m_q\omega^2\lambda_2, \tag{34c}$$

$$\frac{dA}{dt} = \frac{P_A}{m_{cl}}, \tag{34d}$$

$$\frac{dP_A}{dt} = -e^2 m_q A\frac{T(I_\lambda)}{I_\lambda}\lambda_2. \tag{34e}$$

This system associates **a kind of phase-space** to the density matrix (23). The non-linear term $T(I_\lambda)$ in (34e) is easily tractable as a function of I, using (33).

This non-linearity is thus replaced by a dependence upon the initial conditions, via the invariant I_λ (which in turn is fixed by $\hat{\rho}(0)$, i.e. by the initial values of the Lagrange multipliers) and I. Also, from Eqs. (32) and (33) we find

$$I_\lambda = \frac{1}{2\hbar} \ln\left(\frac{\sqrt{I}+\frac{\hbar}{2}}{\sqrt{I}-\frac{\hbar}{2}}\right), \qquad (35)$$

relating I_λ to I. Note here that as I decreases, I_λ augments. If I approaches $\hbar^2/4$, then $I_\lambda \to \infty$, since $X^2 + \hat{P}^2$ approaches the ground state. Even then $I \neq 0$. Thus, we do not reach the classical limit yet. We need to take the limit $\hbar \to 0$ and still $I_\lambda \to \infty$ holds.

7. RESULTS

We obtained the solutions for the system of nonlinear equations (34) for different values of E_r, in the range $[1,\infty]$ [8]. In studies carried out in the past, we have found that the system displays three regions, a quasi-quantum ($E_r \simeq 1$) one, a classic one that starts at E_r^{cl}, and a semiclassical transition zone as in the case of the particular mean values of $(\hat{x}^2, \hat{p}^2, \hat{L} = \hat{x}\hat{p} + \hat{p})$ [8]. Within such an interval we highlight the important value $E_r = E_r^P$, at which chaos emerges [20]. Remind that the system (3) does not explicitly depend upon $\hbar$. EV's depend on it only implicitly, via I, through the initial conditions. Similarly, Eqs. (24) do not depend on $\hbar$. Instead, the system (34) does. The multipliers depend on it via I_λ given by (35), but the normalized quantities $\lambda_i^N = \lambda_i/I_\lambda$, with $i = 1, 2, 3$, do not depend on the actual numerical value of $\hbar$ (see Eqs. (31) and (33)). The curves' morphology does not change with the numerical value of $\hbar$, that has influence only on the graphs' scaling. We are interested in the classical limit only as a function of E_r.

In Figure 1 we depict some relevant Poincaré surfaces (sections' cuts with $A = 0$) corresponding to the system of equations (34). In the graph we plot λ_3^N vs. λ_1^N. Numerical details are given in the pertinent captions. The existence of chaos was verified via the calculation of the Lyapunov characteristic exponents, which are positive for the curves in question. We have also checked on the accuracy of all results by verifying the constancy in time of the dynamical invariants E and I (within a precision of 10^{-10}). Sub-figure (A) corresponds to $E_r \simeq 1$ (quasi-quantal case), displaying periodicity and quasi-periodicity. Sub-figure (B) corresponds to the quantum-classical transition zone, where chaos

Semiquantum Time Evolution II: Density Matrices

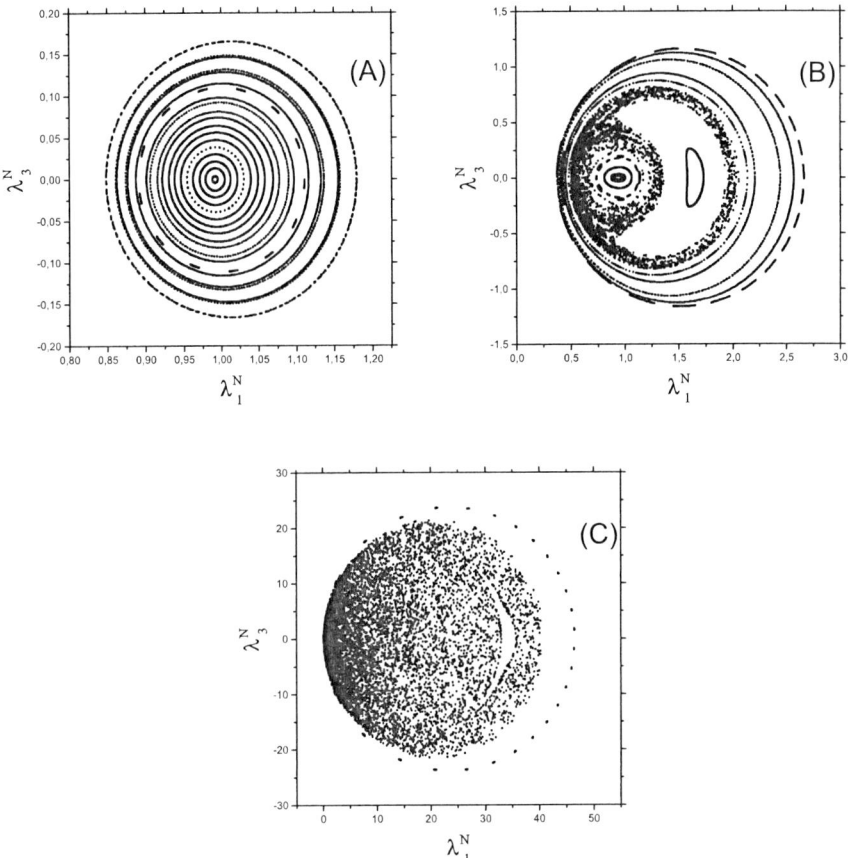

Figure 1. Poincaré surfaces of section: λ_3^N vs. λ_1^N, for $E = 0.6$, $A(t = 0) = 0$, $m_q = m_{cl} = \omega_q = e = 1$. E_r adopts the following values: A) $1,0142$ ("quantum-like" regime), B) $1,5492$ (semi-classical regime), C) $E_r{}^{cl} = 24,2452$ (convergence to the classical limit becomes noticeable).

emerges. Sub-figure (C) corresponds to E_r^{cl}, value of E_r at which the classical zone begins. This type of graphics closely resembles, morphologically, those for the concomitant EV's. This is obvious from Eqs. (31) together with (32). However important changes of scale are registered. Once we have the $\lambda_i(t)$ values at our disposal, we can compute expectation values (EV) as a function of time, that depend upon the λ's via (18). We thus obtain the EVs for the set

($\hat{x}^2$, $\hat{p}^2$, $\hat{L}$). In this case we also can directly use Eqs. (31). In Figure 1 of the chapter *Semiquantum time evolution: Classical limit. Dissipation and quantum measurement* of this book, we depict the corresponding EV's-associated $A = 0$ Poincaré surfaces. We use in this Figure the same parameters and initial conditions. We see the same regions as in Figure 1 of this chapter. Now, in the EV's case, we can register the convergence to the solution of the purely classical system of equations. We also note the coexistence of the Uncertainty Principle with chaos.

Dynamics of the type above described should be replicated for any operator one deals with it via (18). One may thus assert that our "u_λ-space" might be regarded as a **kind of pseudo phase space associated to our density matrix**.

8. DENSITY MATRIX AND A DISSIPATIVE SYSTEM

We consider rhe interaction between a spin or a two level system with a single mode of the electromagnetic field, as in the previous chapter. This is a problem of great interest in several fields, e.g. quantum optics, quantum electronics and magnetic resonance. We consider a two-level Hamiltonian coupled to a classical oscillator (note that dimensionless quantities are employed) ([6])

$$\hat{H} = E_1 \hat{a}_1^\dagger \hat{a}_1 + E_2 \hat{a}_2^\dagger \hat{a}_2 + \frac{\omega}{2}(p_s^2 + s^2) + \gamma s(\epsilon a_1^\dagger \hat{a}_2 + \epsilon^\dagger \hat{a}_2^\dagger \hat{a}_1), \quad (36)$$

were we assume $E_2 > E_1$, define the frequency $\omega_0 = (E_2 - E_1)$ and shall measure the time in terms of $\tau = \omega_0 t$. Here γ is a coupling constant with dimension of energy and ϵ is chosen as a dimensionless parameter, while $\hat{a}_1^\dagger$, $\hat{a}_1$ and $\hat{a}_2^\dagger$, $\hat{a}_2$, are the bosonic creation and annihilation operators of a particle in respectively, levels one and two, s is a classical position variable and p_s the concomitant momentum. We take that N particles are distributed between the two levels.

A partial Lie algebra, given by Eqs. (4), follows if we choose as *relevant* operators those belonging to the set $\{\hat{O}_1 = \hat{a}_1^\dagger \hat{a}_1, \hat{O}_2 = \hat{a}_2^\dagger \hat{a}_2, \hat{O}_3 = i(\epsilon \hat{a}_1^\dagger \hat{a}_2 - \epsilon^\dagger \hat{a}_2^\dagger \hat{a}_1), \hat{O}_4 = (\epsilon \hat{a}_1^\dagger \hat{a}_2 + \epsilon^\dagger \hat{a}_2^\dagger \hat{a}_1)\}$, and also if we choose as *relevant* operators those belonging to the set $\{\Delta \hat{N} = \hat{O}_2 - \hat{O}_1, \hat{O}_3, \hat{O}_4\}$. We assume, at $\tau = 0$, knowlege of the EV's of an appropriate subset of operators, as, for instance,

$\{\Delta\hat{N}, \hat{O}_3, $ y $\hat{O}_4\}$, which suffices to determine the "initial" $\hat{\rho}(\tau = 0)$, whose *form* will be maintained troughout, i.e.

$$\hat{\rho}(\tau) = \exp\left[-\left(\lambda_0 \hat{I} + \lambda_{\Delta N}\Delta\hat{N} + \lambda_3 \hat{O}_3 + \lambda_4 \hat{O}_4\right)\right], \tag{37}$$

where $\lambda_{\Delta N}, \lambda_3$ y λ_4 are (in self-explanatory notation) the relevant Lagrange multipliers entering the present game. Eqs. (17) acquire now the appearance ([6])

$$\frac{d\lambda_{\Delta N}}{d\tau} = \alpha s \lambda_3, \tag{38a}$$

$$\frac{d\lambda_3}{d\tau} = -\alpha s \lambda_{\Delta N} + \lambda_4, \tag{38b}$$

$$\frac{d\lambda_4}{d\tau} = -\lambda_3, \tag{38c}$$

where $\alpha = 2\gamma/\omega_0$ is a useful dimensionless quantity. Of course, our system (38) depends upon the classical variable s, which forces consideration of the Eq. (6b), that involve the parameter η. This equation can be rewritten in the form (for the sake of notational simplicity we shall write p instead of p_s)

$$\frac{ds}{d\tau} = \Omega p, \tag{39a}$$

$$\frac{dp}{d\tau} = -(\Omega s + \frac{1}{2}\alpha\langle\hat{O}_4\rangle + \delta p), \tag{39b}$$

where we have introduced the dimensionless parameters ([6])

$$\Omega = \frac{\omega}{\omega_0}, \qquad \delta = \frac{\eta}{\omega_0}. \tag{40}$$

The presence of the term $\langle\hat{O}_4\rangle$ in (39b), introduces (in principle) a non-linearity (as such a term is a function of the multipliers) through

$$\langle\hat{O}_4\rangle(\tau) = \text{Tr}[\hat{\rho}(\tau)\hat{O}_4], \tag{41}$$

but we will presently see that this non-linearity can be easily dealt with.

9. GENERALIZED BLOCH EQUATIONS

Notice that the quantity I_λ defined as

$$I_\lambda = \left(\lambda_{\Delta N}{}^2 + \lambda_3{}^2 + \lambda_4{}^2\right)^{1/2}, \tag{42}$$

is, by virtue of (38) a constant of the motion, i.e. $dI_\lambda/dt = 0$. After a little algebra, it is easy to see that, using Eq. (13) (or using $\langle \hat{O} \rangle(\tau) = \text{Tr}[\hat{\rho}(\tau)\hat{O}]$), the particular EV's of, respectively, $\Delta \hat{N}, \hat{O}_3,$ y $\hat{O}_4$ can be written in the fashion

$$\langle \Delta \hat{N} \rangle = \lambda_{\Delta N} \frac{T(I_\lambda)}{I_\lambda}, \tag{43a}$$

$$\langle \hat{O}_3 \rangle = \lambda_3 \frac{T(I_\lambda)}{I_\lambda}, \tag{43b}$$

$$\langle \hat{O}_4 \rangle = \lambda_4 \frac{T(I_\lambda)}{I_\lambda}, \tag{43c}$$

with $T(I_\lambda)$ given by

$$T(I_\lambda) = \frac{\sum_{l=0}^{N}(N-2l)\exp[-(N-2l)I_\lambda]}{\sum_{l=0}^{N}\exp[-(N-2l)I_\lambda]}. \tag{44}$$

Further, we deduce from (43) ($T(I_\lambda) < 0$) that

$$T(I_\lambda) = -I, \tag{45}$$

where I is the so-called Bloch vector "length"

$$I = \left(\Delta \hat{N}^2 + \langle \hat{O}_3 \rangle^2 + \langle \hat{O}_4 \rangle^2\right)^{1/2}. \tag{46}$$

Now we are in a position to conveniently recast our dynamical system in "u_λ-space" by recourse to (43c), eliminating the non-linearity referred to above, and obtain a tractable system of equations. We have

$$\frac{d\lambda_{\Delta N}}{d\tau} = \alpha s \lambda_3, \tag{47a}$$

$$\frac{d\lambda_3}{d\tau} = -\alpha s \lambda_{\Delta N} + \lambda_4, \tag{47b}$$

$$\frac{d\lambda_4}{d\tau} = -\lambda_3, \tag{47c}$$

$$\frac{ds}{d\tau} = \Omega p, \tag{47d}$$

$$\frac{dp}{d\tau} = -\left(\Omega s + \frac{1}{2}\alpha\lambda_4\frac{T(I_\lambda)}{I_\lambda} + \delta p\right), \tag{47e}$$

which constitutes a set of Bloch-like equations in "u_λ-space".

The non-linear term $T(I_\lambda)/I_\lambda$ in (47e) is easily removed, since I_λ is a constant of the motion, and so are also $T(I_\lambda)$ and I. The non-linearity is thus replaced by a dependence of our set of equations upon the initial conditions, via the invariant I_λ (which in turn is fixed by $\hat\rho(0)$, i.e. by the initial values of the Lagrange multipliers).

9.1. The Fixed Points

Our autonomous system is now easily solved (see Appendix A). Elementary considerations allow then to determine the fixed points of this system. For our generalized vector $\vec u_\lambda$, whose components are Lagrange multipliers *and* classical variables, the fixed points $(\lambda_{\Delta Nf}, \lambda_{3f}, \lambda_{4f}, s_f, p_f)$ of (47) are of *two* kinds: i) type A, for which the component s_f remains *finite*, and ii) type B, for which this component vanishes.

For type A we have ([6])

$$\lambda_{\Delta Nf} = -2\frac{\Omega}{\alpha^2}\frac{I_\lambda}{T(I_\lambda)}, \tag{48a}$$

$$\lambda_{3f} = 0, \tag{48b}$$

$$\lambda_{4f} = \pm\left(I_\lambda^2 - 4\frac{\Omega^2}{\alpha^4}\frac{I_\lambda^2}{T(I_\lambda)^2}\right)^{1/2}, \tag{48c}$$

$$s_f = -\frac{\alpha}{2\Omega}\lambda_4\frac{T(I_\lambda)}{I_\lambda}, \tag{48d}$$

$$p_f = 0, \tag{48e}$$

results that hold for $2\Omega/\alpha^2 < -T(I_\lambda)$. If this is not the case, the fixed points are of type B with

$$\lambda_{\Delta Nf} = \pm I_\lambda, \quad (49a)$$
$$\lambda_{3f} = 0, \quad (49b)$$
$$\lambda_{4f} = 0, \quad (49c)$$
$$s_f = 0, \quad (49d)$$
$$p_f = 0. \quad (49e)$$

The fixed points are seen to depend upon both the initial conditions (through the invariant I_λ) and the parameters α and Ω. An abrupt change of behaviour ensues at

$$\frac{2\Omega}{\alpha^2} = -T(I_\lambda). \quad (50)$$

It is worth-while to point out that the dissipative parameter δ does not influence the nature of the fixed points. In other words, this nature is only affected by the interaction between the classical and the quantum system. *The attractors exists because δ is not zero, but their location is independent of the precise value that δ may adopt.*

The Lagrange multipliers $\lambda_{\Delta N}$, λ_3, and λ_4 "move" on the surface of a sphere of radius I_λ, in that three-dimensional subspace determined by the projection of the solutions of (47) upon the "λ-space". Figure 1 of reference [6], depicts some typical results for the temporal evolution of our Lagrange multipliers.

9.2. Expectation Values

In order to evaluate EV's it is convenient to perform a change of representation by recourse to a Bogoliubov-like transformation (BT)

$$\hat{b}_1 = (\cos\theta\cos\sigma + i\sin\theta\sin\sigma)\hat{a}_1 - (\sin\theta\cos\sigma - i\cos\theta\sin\sigma)\hat{a}_2 \quad (51a)$$
$$\hat{b}_2 = (\sin\theta\cos\sigma + i\cos\theta\sin\sigma)\hat{a}_1 + (\cos\theta\cos\sigma - i\sin\theta\sin\sigma)\hat{a}_2 \quad (51b)$$

where the operators $\hat{b}_1$ and $\hat{b}_2$ verify $[\hat{b}_1, \hat{b}_1^\dagger] = 1$ and $[\hat{b}_2, \hat{b}_2^\dagger] = 1$, with $\hat{b}_1^\dagger$ y $\hat{b}_2^\dagger$ the hermitians conjugates of $\hat{b}_1$ y $\hat{b}_2$, respectively. The angles θ y σ are determined according to

$$\sin 2\theta = \frac{\lambda_4}{(\lambda_{\Delta N}^2 + \lambda_4^2)^{1/2}}, \tag{52a}$$

$$\cos 2\theta = \frac{\lambda_{\Delta N}}{(\lambda_{\Delta N}^2 + \lambda_4^2)^{1/2}}, \tag{52b}$$

and

$$\sin 2\sigma = -\frac{\lambda_3}{I_\lambda}, \tag{53a}$$

$$\cos 2\sigma = \frac{1}{I_\lambda}(\lambda_{\Delta N}^2 + \lambda_4^2)^{1/2}. \tag{53b}$$

The BT allows for a simpler $\hat{\rho}$-form

$$\hat{\rho}(\tau) = \exp\left(-\lambda_0 \hat{I} - I_\lambda \Delta \hat{N}^b\right), \tag{54}$$

with $\Delta \hat{N}^b$ given by

$$\Delta \hat{N}^b = \hat{O}_2^b - \hat{O}_1^b, \tag{55}$$

where the sum of $\hat{O}_1^b = \hat{b}_1^\dagger \hat{b}_1$ and $\hat{O}_2 = \hat{b}_2^\dagger \hat{b}_2$ satisfies

$$\langle \hat{O}_1^b \rangle + \langle \hat{O}_2^b \rangle = N, \tag{56}$$

since the BT conserves the particle-number N.

Our main interest here refers to the limit $\tau \to \infty$. Since $\lambda_{3f} = 0$ for both kinds of fixed points (A and B), we find (cf. Eq. (52)) $\sigma_f = 0$, (of course, the subindex f indicates in all cases that one is talking about the fixed points (48) or (49)). For the other relevant angle we have (cf. Eq. (53))

$$\sin 2\theta_f = \frac{\lambda_{4f}}{(\lambda_{\Delta Nf}^2 + \lambda_{4f}^2)^{1/2}}, \tag{57a}$$

$$\cos 2\theta_f = \frac{\lambda_{\Delta Nf}}{(\lambda_{\Delta Nf}^2 + \lambda_{4f}^2)^{1/2}}, \tag{57b}$$

which allows for a simplification of the BT (51), that acquires now the aspect

$$\hat{b}_1 = \cos\theta_f \hat{a}_1 - \sin\theta_f \hat{a}_2, \tag{58a}$$

$$\hat{b}_2 = \sin\theta_f \hat{a}_1 + \cos\theta_f \hat{a}_2. \tag{58b}$$

For operators of the form

$$\hat{O}^{n,m,n',m'} = (\hat{a}_1)^{\dagger n}(\hat{a}_2^\dagger)^m (\hat{a}_1)^{n'} (\hat{a}_2)^{m'}, \quad (59)$$

utilizing Eqs. (58) and (18) one finds that their EV *for fixed points of type A* $(\tau \to \infty)$ is

$$\langle \hat{O}^{n,m,n',m'} \rangle_f = 0, \quad (60)$$

for $m + n \neq m' + n'$. If, instead, $m + n = m' + n'$ we obtain

$$\langle \hat{O}^{n,m,n',m'} \rangle_f = \sum_{i=0}^{n}\sum_{j=0}^{m}\sum_{k=0}^{n'}(-1)^{(i+2j-k)} C_i{}^n C_j{}^m C_k{}^{n'}$$
$$(\cos\theta_f)^{(n+2m-n')}(\sin\theta_f)^{(n+n'+2i-2k)} \langle \hat{O}_b^{n,m,i,j} \rangle, \quad (61)$$

where $C_i{}^n$, $C_j{}^m$, y $C_k{}^{n'}$ are appropriate combinatorial numbers and $\langle \hat{O}_b^{n,m,i,j} \rangle$ stands for the EV of

$$\hat{O}_b^{n,m,i,j} = (\hat{b}_1^\dagger \hat{b}_1)^{(i+j)} (\hat{b}_2^\dagger \hat{b}_2)^{(n+m-i-j)}, \quad (62)$$

being given by

$$\langle \hat{O}_b^{n,m,i,j} \rangle = \frac{\sum_{l=0}^{N} l^{i+j}(N-l)^{m+n-i-j}\exp[-(N-2l)I_\lambda]}{\sum_{l=0}^{N}\exp[-(N-2l)I_\lambda]}, \quad (63)$$

which is seen to be a constant of the motion.

In the case of EV's of hermitian operators of the type $\hat{O}_+ = (\hat{O} + \hat{O}^\dagger)/2$ and $\hat{O}_- = (\hat{O} - \hat{O}^\dagger)/2i$ one finds

$$\langle \hat{O}_+ \rangle_f = \langle \hat{O} \rangle_f, \quad (64a)$$
$$\langle \hat{O}_- \rangle_f = 0. \quad (64b)$$

Things become simpler *for fixed points of type B*, as $\lambda_{3f} = 0$ and $\lambda_{4f} = 0$. Thus, $\theta_f = 0$ and no BT transformation is needed. For (59) we have

$$\langle \hat{O} \rangle_f = 0, \quad (65)$$

if $n \neq n'$ or $m \neq m'$. For the diagonal case one finds

$$\langle \hat{O} \rangle_f = \frac{\sum_{l=0}^{N} l^n (N-l)^m \exp[-(N-2l)I_\lambda]}{\sum_{l=0}^{N} \exp[-(N-2l)I_\lambda]}. \quad (66)$$

In the case of the operators $\hat{O}_+$ and $\hat{O}_-$, the results (64a) and (64b) also hold, with $\langle \hat{O} \rangle_f$ given by (66). As special instances we have, for $\hat{\Delta N}$, $\hat{O}_3$, and $\hat{O}_4$ results similar to those of (48) and (49), namely, for type A

$$\langle \Delta \hat{N} \rangle_f = -2\frac{\Omega}{\alpha^2}, \quad (67a)$$

$$\langle \hat{O}_3 \rangle_f = 0, \quad (67b)$$

$$\langle \hat{O}_4 \rangle_f = \pm (I^2 - 4\frac{\Omega^2}{\alpha^4})^{1/2}, \quad (67c)$$

valid for $2\Omega/\alpha^2 < I$. Otherwise on is lead to the type B results

$$\langle \Delta \hat{N} \rangle_f = \pm I, \quad (68a)$$

$$\langle \hat{O}_3 \rangle_f = 0, \quad (68b)$$

$$\langle \hat{O}_4 \rangle_f = 0. \quad (68c)$$

Another special case worth mentioning is that of the following EV's:

$$\langle (\hat{a}_1^\dagger)^n \rangle_f = 0, \quad (69a)$$

$$\langle (\hat{a}_2^\dagger)^n \rangle_f = 0, \quad (69b)$$

$$\langle (\hat{a}_1)^n \rangle_f = 0, \quad (69c)$$

$$\langle (\hat{a}_2)^n \rangle_f = 0. \quad (69d)$$

9.3. Stability Considerations

It becomes mandatory now to study the stability of our equilibrium EV's, which is related to that of the de Lagrange multipliers: small variations in the (initial) EV's of $\Delta \hat{N}$, $\hat{O}_3$, y $\hat{O}_4$ imply small variations in the associated λ_i's (and viceversa, see Eqs. (43) and (44)), as those EV's determine (at the initial time) the statistical operator (37), and, a posteriori, all EV's at that time. In order to discuss the stability of the fixed points we follow the usual route: linearize (47) around the fixed points and find the eigenvalues of the associated matrix (see Appendix A).

The stable points are, for type A- instances (i.e. for $2\Omega/\alpha^2 < -T(I_\lambda)$)

$$\lambda_{\Delta N f} = -2\frac{\Omega}{\alpha^2}\frac{I_\lambda}{T(I_\lambda)}, \tag{70a}$$

$$\lambda_{3f} = 0, \tag{70b}$$

$$\lambda_{4f} = \pm\left(I_\lambda^2 - 4\frac{\Omega^2}{\alpha^4}\frac{I_\lambda^2}{T(I_\lambda)^2}\right)^{1/2}, \tag{70c}$$

$$s_f = -\frac{\alpha}{2\Omega}\lambda_4\frac{T(I_\lambda)}{I_\lambda}, \tag{70d}$$

$$p_f = 0, \tag{70e}$$

while, for type B situations (i.e. for $2\Omega/\alpha^2 \geq -T(I_\lambda)$) one has

$$\lambda_{\Delta N f} = I_\lambda, \tag{71a}$$

$$\lambda_{3f} = 0, \tag{71b}$$

$$\lambda_{4f} = 0, \tag{71c}$$

$$s_f = 0, \tag{71d}$$

$$p_f = 0. \tag{71e}$$

After a detailed numerical study that entails variations of both i) initial conditions and ii) values of the model-parameters, we find that the fixed points (70) and (71) are indeed attractors ([6]).

It is of interest to mention that for the entropy (15) one obtains

$$S = \lambda_0 - I_\lambda I, \tag{72}$$

i.e., S is a function of the invariants I_λ and I.

Summing up, no matter where we start (values of the parameters of the model, initial conditions), the system represented by the hamiltonian (36) ends up in an equilibrium situation appropriately described by either

$$\hat{\rho}(\tau) = \exp\left[-\left(\lambda_0 \hat{I} + \lambda_{\Delta N f}\Delta\hat{N} + \lambda_{4f}\hat{O}_4\right)\right], \tag{73}$$

(for $2\Omega/\alpha^2 < -T(I_\lambda)$), or by

$$\hat{\rho}(\tau) = \exp\left[-\left(\lambda_0 \hat{I} + \lambda_{\Delta N f}\Delta\hat{N}\right)\right], \tag{74}$$

(for $2\Omega/\alpha^2 \geq -T(I_\lambda)$). The associated equilibrium (or "final") Lagrange multipliers are given above (cf. Eqs. (70) and (71)).

During the equilibration process the purely quantal entropy, given in terms of $\hat{\rho}$, remains constant. Of course, this entropy must be supplemented, in order to have the total one, by those of both the classical system and the reservoir. But the details are unimportant: what matters is that a purely quantal evolution (constant quantal entropy) has been obtained that, without violation of any quantum rule, is indeed accompanied by dissipation.

10. LINEARIZATION PROCEDURE

The stability of the fixed points is determined by linearizing (47) around the fixed points, i.e., we insert

$$(\lambda_{\Delta N}, \lambda_3, \lambda_4, s, p) = (\lambda_{\Delta N f} + \epsilon_{\Delta N}, \lambda_{3f} + \epsilon_3, \lambda_{4f} + \epsilon_4, s_f + \epsilon_s, p_f + \epsilon_p), \quad (75)$$

in those equations, which leads, for both pertinent types, A and B, to the following system of equations

$$\frac{d\epsilon_{\Delta N}}{d\tau} = \alpha s_f \epsilon_3, \tag{76a}$$

$$\frac{d\epsilon_3}{d\tau} = -\alpha \left(s_f \epsilon_{\Delta N} + \lambda_{\Delta N f} \epsilon_s \right) + \epsilon_4, \tag{76b}$$

$$\frac{d\epsilon_4}{d\tau} = -\epsilon_3, \tag{76c}$$

$$\frac{d\epsilon_s}{d\tau} = \Omega \epsilon_p, \tag{76d}$$

$$\frac{d\epsilon_p}{d\tau} = -\left(\mu \epsilon_{\Delta N} + \nu \epsilon_4 + \Omega \epsilon_s + \delta \epsilon_p \right), \tag{76e}$$

where

$$\mu = \frac{1}{2} \alpha \lambda_{4f} \lambda_{\Delta N f} A(I_{\lambda f}), \tag{77a}$$

$$\nu = \frac{1}{2} \alpha \left(\lambda_{4f}^2 A(I_{\lambda f}) + \frac{T((I_{\lambda f}))}{(I_{\lambda f})} \right), \tag{77b}$$

$$A(I_{\lambda f}) = \frac{1}{I_{\lambda f}^2} \left(\frac{dT}{dI_\lambda}(I_{\lambda f}) - \frac{T(I_{\lambda f})}{I_{\lambda f}} \right). \tag{77c}$$

Notice that we have used in Eq. (76e) the fact that, according to (75), the quantity $T(I_\lambda{}^n)/I_\lambda{}^n$ may be written in the fashion

$$\frac{T(I_\lambda{}^n)}{I_\lambda{}^n} \simeq \frac{T(I_{\lambda f})}{I_{\lambda f}} + \frac{1}{2} A(I_{\lambda f}) I_\lambda(\epsilon), \tag{78}$$

where $I_\lambda{}^n$ and $I_{\lambda f}$ are

$$I_\lambda{}^n = \left[\left(\lambda_{\Delta Nf} + \epsilon_{\Delta N}(0)\right)^2 + \left(\lambda_{3f} + \epsilon_3(0)\right)^2 + \left(\lambda_{4f} + \epsilon_4(0)\right)^2 \right]^{1/2}, \tag{79a}$$

$$I_{\lambda f} = (\lambda_{\Delta Nf}{}^2 + \lambda_{3f}{}^2 + \lambda_{4f}{}^2)^{1/2}, \tag{79b}$$

and where

$$I_\lambda(\epsilon) = 2 \left(\lambda_{\Delta Nf} \epsilon_{\Delta N}(0) + \lambda_{4f} \epsilon_4(0)\right), \tag{80}$$

is the invariant of the motion corresponding to Eqs. (76).

For type A the eigenvalues of the concomitant secular matrix are the roots of either

$$r = 0, \tag{81}$$

or, alternatively,

$$r^4 + \delta r^3 + (\Omega^2 + 1 + \alpha^2 s_f^2) r^2 + \delta (1 + \alpha^2 s_f^2) r + \Omega^2 \alpha^2 s_f^2 = 0, \tag{82}$$

while for type B we find either

$$r = 0, \tag{83}$$

or, alternatively,

$$r^4 + \delta r^3 + (\Omega^2 + 1) r^2 + \delta r + \Omega^2 + \frac{\alpha^2}{2} \Omega \lambda_{\Delta Nf} \frac{T(I_{\lambda f})}{I_{\lambda f}} = 0. \tag{84}$$

In general, the roots of Eqs. (82) and (84) can be catalogued in the following fashion: i) two pairs of complex conjugate roots, ii) two real roots together with a pair of complex conjugate ones, iii) four real roots. The real part of the complex conjugate roots does not vanish. For type A one obtains either negative real roots or complex real roots with negative real parts after solving Eq. (82). The solutions of Eqs. (76) evolve, for $\tau \to \infty$, towards values consistent with a

Semiquantum Time Evolution II: Density Matrices

root $r = 0$ (cf. Eq. (81)). One finds

$$\epsilon_{\Delta Nf} = -\frac{1}{2}\lambda_{\Delta Nf}\frac{I_{\lambda f}}{T(I_{\lambda f})}A(I_{\lambda f})I_{\lambda}(\epsilon), \quad (85a)$$

$$\epsilon_{3f} = \epsilon_{pf} = 0, \quad (85b)$$

$$\epsilon_{sf} = -\frac{\alpha}{2\Omega}\left(A(I_{\lambda f})\lambda_{4f}\frac{I_{\lambda}(\epsilon)}{2} + \frac{T(I_{\lambda f})}{I_{\lambda f}}\epsilon_{4f}\right). \quad (85c)$$

The value of ϵ_{4f} (in Eq. (85c)) is determined using $I_{\lambda}(\epsilon)$, which yields

$$\epsilon_{4f} = \frac{1}{2\lambda_{4f}}I_{\lambda}(\epsilon)\left(1 + A(I_{\lambda f})(\lambda_{\Delta Nf})^2\frac{I_{\lambda f}}{2T(I_{\lambda f})}\right). \quad (86)$$

Therefore, trajectories starting near the fixed points (48) move toward neighbouring fixed points (denoted with a superindex n), given by

$$\lambda_{\Delta Nf}^n = \lambda_{\Delta Nf}\left(1 - \frac{1}{2}\frac{I_{\lambda f}}{T(I_{\lambda f})}A(I_{\lambda f})I_{\lambda}(\epsilon)\right), \quad (87a)$$

$$\lambda_{3f}^n = \lambda_{3f}, \quad (87b)$$

$$\lambda_{4f}^n = \lambda_{4f} + \frac{1}{2\lambda_{4f}}I_{\lambda}(\epsilon)\left(1 + A(I_{\lambda f})(\lambda_{\Delta Nf})2\frac{I_{\lambda f}}{2T(I_{\lambda f})}\right), \quad (87c)$$

$$s_f^n = s_f - \frac{\alpha}{2\Omega}\left(A(I_{\lambda f})\lambda_{4f}\frac{I_{\lambda}(\epsilon)}{2} + \frac{T(I_{\lambda f})}{I_{\lambda f}}\epsilon_{4f}\right), \quad (87d)$$

$$p_f^n = p_f. \quad (87e)$$

Negative real roots and negative real parts of the complex roots are obtained for Eq. (84) if, in Eq. (49a),

$$\lambda_{\Delta Nf} = I_{\lambda}, \quad (88)$$

and, in addition, the relation $2\Omega/\alpha^2 \geq -T(I_{\lambda})$ is satisfied. As a consequence, the solutions of Eqs. (76) evolve, for $\tau \to \infty$, to values corresponding to the eigenvalue zero (Eq. (83)), as in case A. The fixed points of Eqs. (76) are determined by Eq. (80), and one finds

$$\epsilon_{\Delta Nf} = \frac{I_{\lambda}(\epsilon)}{2\Delta N_f}, \quad (89a)$$

$$\epsilon_{3f} = \epsilon_{4f} = \epsilon_{sf} = \epsilon_{pf} = 0, \quad (89b)$$

so that one obtains

$$\lambda_{\Delta N_f^n} = \lambda_{\Delta N f} + \frac{I_\lambda(\epsilon)}{2\lambda_{\Delta N f}}, \tag{90a}$$

$$\lambda_{3f}^n = \lambda_{3f}, \tag{90b}$$

$$\lambda_{4f}^n = \lambda_{4f}, \tag{90c}$$

$$s_f^n = s_f, \tag{90d}$$

$$p_f^n = p_f. \tag{90e}$$

In general, trajectories starting in the vecinity of the fixed points move toward neighbouring fixed points (for both types A and B). The concomitant shift, as it is easy to see (cf. Eqs. (87) and (90)), is proportional to the perturbation. In particular, if the perturbation does not modify the temporal invariant I_λ, the trajectories lead back to the original fixed points, i.e.

$$\begin{aligned} I_\lambda^n &\simeq I_{\lambda f} + I_\lambda(\epsilon) \\ &= I_{\lambda f}, \end{aligned} \tag{91}$$

so that $I_\lambda(\epsilon) = 0$ (we have neglected powers of ϵ higher than the linear one and remembered that in both cases λ_{3f} vanishes). Therefore, if we start in the neighbourhood of the fixed points we either return to them or remain in their vecinity, i.e., the fixed points are stable ones [2].

On the other hand, if the opposite sign in (49a) is chosen, complex roots with positive real parts ensue, so that the fixed points turn out to be unstable ones.

Acknowledgments

A. M. K. fully acknowledges support from the Comisión de Investigaciones Científicas de la Provincia de Buenos Aires (CICPBA).

Conclusion

In this chapter we have presented a methodology of our own for obtaining the MaxEnt density matrix (MDM) describing the semiquantum time evolution. Its, for all times, is given by Eq. (11). This MDM always satifies the Liouville equation. Of course, this is so if the operators entering tha MDM happen to satisfy a (commutation) partial Lie algebra with the total Hamiltonian. The time-evolution is given in term of the MaxEnt Lagrange multipliers, that comply with the system od Eqs. (19), that involves classical variables. For $\eta = 0$ we face a conservative instance and for and $\eta > 0$ we confront a dissipative one. Mean values at any time are given by Eq. (18).

Two different physical examples have been examined that involve matter interacting with an EM field. In one of them we have investigated the classical limit of quantum mechanics and highlighted the coexistence of quantum uncertainty with deterministic chaos. In the second example we studied quantum friction, also coexisting with uncertainty.

Summing up, we have showed just how chaos and dissipation can happily coexist with Heisenberg's uncertainty and with a strict Liouville's time evolution.

REFERENCES

[1] Bloch, E. (1946). Nuclear Induction. *Phys. Rev.* 70, 460.

[2] Milonni, P., Shih, M., Ackerhalt, J.R. (1987). *Chaos in Laser-Matter Interactions* (World Scientific Publishing Co.: Singapore).

[3] Meystre, P., Sargent, M. (1991). *Elements of Quantum Optics* (Springer, NY).

[4] Ring, P., Schuck, P. (1980). *The Nuclear Many-Body Problem* (Springer-Verlag: Berlin, Germany).

[5] Kowalski, A.M., Plastino, A and Proto, A.N. (1995). Semiclassical model for quantum dissipation. *Phys. Rev. E* 52, 165.

[6] Kowalski, A.M., Plastino, A and Proto, A.N. (1997). A semiclassical statistical model for quantum dissipation. *Physica A* 236, 429.

[7] Kowalski, A.M., Martin, M.T., Nuñez, J., Plastino, A and Proto, A.N. (1998). Quantitative indicator for semi-quantum chaos. *Phys. Rev. A* 58, 2596.

[8] Kowalski, A.M., Plastino, A and Proto, A.N. (2002). Classical limits. *Phys. Lett.* A 297, 162.

[9] Kowalski, A.M., R. Rossignoli, R. (2018). Nonlinear dynamics of a semi-quantum Hamiltonian in the vicinity of quantum unstable regimes. *Chaos, Solitons and Fractals* 109, 140.

[10] Kowalski, A.M., Plastino, A., Rossignoli, R. (2019). Complexity of a matter-field Hamiltonian in the vicinity of a quantum instability. *Physica A* 513, 767.

[11] Kowalski, A.M., Plastino, A. (2019). A nonlinear matter-field Hamiltonian analyzed with Renyi and Tsallis statistics. *Physica A* 535, 122387.

[12] Kowalski, A.M. (2016). Betting on dynamics. *Physica A* 458, 106.

[13] Halliwell, J.J., Yearsley, J.M. (2009). Arrival times, complex potentials, and decoherent histories. *Phys. Rev. A* 79, 062101:1.

[14] Everitt, M.J., Munro, W.J, Spiller, T.P. (2009). Quantum-classical crossover of a field mode. *Phys. Rev. A* 79, 032328:1.

[15] Zeh, H.D. (1999). Why Bohms quantum theory?. *Found. Phys. Lett.* 12, 197.

[16] Zurek, W.H. (1981). Pointer basis of quantum apparatus: Into what mixture does the wave packet collapse?. *Phys. Rev. D* 24, 1516.

[17] Zurek, W.H. (2003). Decoherence, einselection, and the quantum origins of the classical. *Rev. Mod. Phys* 75, 715.

[18] Cooper, F., Dawson, J., Habib, S., Ryne, R.D. (1998). Chaos in time-dependent variational approximations to quantum dynamics. *Phys. Rev. E* 57, 1489.

[19] Kowalski, A.M., Martín, M.T., Plastino, A. Rosso, O.A., (2007). Bandt-Pompe approach to the classical-quantum transition. *Phys. D* 233, 21.

[20] Kowalski, A.M., Plastino, A. (2009). Bandt-Pompe-Tsallis quantifier and quantum-classical transition. *Physica A* 388, 4061.

[21] Kowalski, A.M., Martín, M.T., Plastino, A. and Judge, G. (2014). Kullback-Leibler approach to chaotic time series. *Transactions on Theoretical Physics* 1, 40.

[22] Kowalski, A.M., Martín, M.T., Plastino, A. (2015). Generalized relative entropies in the classical limit. *Physica A* 422, 167.

[23] Weiss U. (2008). *Quantum Dissipative Systems* (Series in Modern Condensed Matter Physics) (Singapore: World Scientific).

[24] Kanai, E. (1948). On the Quantization of the Dissipative Systems. *Prog. Theor. Phys.* 3, 440.

[25] Hasse, R.W. (1978). Microscopic derivation of quantum fluctuations in nuclear reactions. *J. of Phys. A* 11, 1245.

[26] Öttinger, H.C. (2011). The geometry and thermodynamics of dissipative quantum systems Hans Christian. *EPL* 94, 10006.

[27] Katz, A. (1967). *Principles of statistical mechanics*. Freeman, San Francisco.

[28] Alhasid, Y, Levine, R.D. (1977). Entropy and Chemical Change. III. The Maximal Entropy (subject to constraints) Procedure as a Dynamical Theory. *J. Chem. Phys.* 67, 4321.

[29] Alhasid, Y, Levine, R.D. (1978). Connection Between the Maximal Entropy and The Scattering Theoretic Analysis of Collision Processes. *Phys. Rev. A* 18, 189.

[30] Arnold, V. I. (1978). *Mathematical methods of classical mechanics*. Springer, Heidelberg-New York.

In: Understanding Time Evolution
Editor: Asger S. Thorsen

ISBN: 978-1-53617-874-6
© 2020 Nova Science Publishers, Inc.

Chapter 3

OBJECTIVE AND NONOBJECTIVE MATHEMATICAL DESCRIPTION OF THE ELECTRIC CHARGE TRANSPORT

Agneta M. Balint[1] *and Stefan Balint*[2,*]

[1]Department of Physics, West University of Timisoara, Timisoara, Romania
[2]Department of Computer Science, West University of Timisoara, Timisoara, Romania

ABSTRACT

In this chapter the objectivity of the mathematical description of the electric charge transport is discussed. It is shown that the description of the electric charge transport in electric circuit, across passive and active biological neuron membrane, along passive and active neuron axons and dendrits and in biological neuron networks, using Caputo or Riemann-Liouville fractional order derivatives defined with integral representation on finite interval, is nonobjective. It is shown also that the mathematical description of these phenomena using integer order derivatives, general Liouville-Caputo or general Riemann-Liouville fractional order derivatives defined with integral representation on infinite interval, is objective.

---

[*] Corresponding Author's Email: stefan.balint@e-uvt.ro (Corresponding author).

Keywords: objectivity of a mathematical description, electric charge transport description, fractional order derivative

INTRODUCTION

Objectivity in science means that qualitative and quantitative descriptions of phenomena remain unchanged when the phenomena are observed by different observers; that is, it is possible to reconcile observations of the process into a single coherent description of it [1].

DESCRIPTION OF THE ELECTRON TRANSPORT IN ELECTRIC CIRCUIT WITH INTEGER ORDER DERIVATIVES IS OBJECTIVE

Observer O describes the electron transport in a classical RLC series circuit with a real valued and real variable function $i = i(t_M)$ [2]. This function represents the variation of the electron flow intensity (current intensity) in the circuit. More exactly, the observer O fixes a moment of time M_O for fixing the origin of the time measurement (for instant the moment of the start of his chronometer) and a unit [second] for time measuring. A moment of time M which is earlier than M_O is represented by a negative real number $t_M < 0$, a moment of time M which is later than M_O is represented by a positive real number $t_M > 0$ and the moment of time M_O is represented by the real number $t_{M_O} = 0$. At any moment of time M, represented by the number t_M, the observer measures the current intensity $i(t_M)$ and describes the electron transport with the function $i = i(t_M)$.

Observer $O*$ uses a similar procedure. For observer $O*$ the origin of the time measurement is M_{O*}, the unit is [second]. A moment of time M which is earlier than M_{O*} for the observer $O*$ is represented by a negative number $t*_M < 0$, the moment of time M_{O*} is represented by the real number

Objective and Nonobjective Mathematical Description ...   57

$t*_{M_O*} = 0$ and a moment of time M which is later than the moment of time M_{O*} is represented by a positive real number $t*_M > 0$. At any moment of time M, represented by $t*_M$, observer $O*$ measures the current intensity $i*(t*_M)$ and describes the transport with the function $i* = i*(t*_M)$. Remark that a moment of time M in case of the observer O is described by the real number t_M and in case of the observer $O*$ by the real number $t*_M$. The numbers t_M and $t*_M$ verify the following relations:

$$t_M = t*_M + t_{M_O*} \tag{1}$$

$$t*_M = t_M + t*_{M_O} \tag{2}$$

In (1) t_{M_O*} is the real number which represents the moment of time M_{O*} in the system of time measuring of observer O and $t*_{M_O}$ in (2) is the real number which represents the moment M_O in the system of time measuring of observer $O*$.

At an arbitrary moment of time M, $i(t_M)$ represents the current intensity in the RLC series circuit measured by the observer O and $i*(t*_M)$ represents the current intensity in the same circuit measured by the observer $O*$. Because the current intensity concerns the same circuit the following relations hold:

$$i(t_M) = i*(t*_M) = i(t*_M + t_{M_O*}) \tag{3}$$

$$i*(t*_M) = i(t_M) = i*(t_M + t*_{M_O}) \tag{4}$$

Relations (3) or (4) reconcile the description made by the two observers, and make possible the description of the electron transport in a classical RLC series circuit by one of the functions $i = i(t_M)$ or $i* = i*(t*_M)$.

By using Ohm's law, Faraday's law, Kirchhoff's current law and Kirchhoff's voltage law it can be shown (see [3]) that the current intensity variation described by the observer O with the function $i = i(t_M)$ verifies the following second order differential equation:

$$L \cdot \frac{d^2 i}{dt_M^2} + R \cdot \frac{di}{dt_M} + \frac{i}{C} = \frac{dV}{dt_M} \qquad (5)$$

Here: L- inductance, R- resistance, C- capacitance and $V = V(t_M)$ describes the variation of the external source voltage in terms of the observer O.

In terms of the description of observer O^*, the same laws lead to the conclusion that the function $i^*(t^*_M)$ verifies the following second order differential equation:

$$L \cdot \frac{d^2 i^*}{dt^{*2}_M} + R \cdot \frac{di^*}{dt^*_M} + \frac{i^*}{C} = \frac{dV^*}{dt^*_M} \qquad (6)$$

In (6) L, R, C are the same constants as in (5) and $V^* = V^*(t^*_M)$ describes the variation of the external source voltage in terms of the observer O^*.

The descriptions $V = V(t_M)$ and $V^* = V^*(t^*_M)$ concern the same source so, for them the following equalities hold:

$$V(t_M) = V^*(t^*_M) = V(t^*_M + t_{M_O^*}) = V^*(t_M + t^*_{M_O}) \qquad (7)$$

Equations (5) and (6) are different, but their solutions describe the current intensity variations in the same classical RLC series circuit under the action of the same external source voltage. This can be proven showing that if $i = i(t_M)$ is a solution of (5), then $i^* = i^*(t^*_M)$, defined by (3), is a solution of (6) and if $i^* = i^*(t^*_M)$ is a solution of (6), then $i = i(t_M)$, defined by (4), is a solution of (5). In other words, the dynamic of the current intensity in the considered

Objective and Nonobjective Mathematical Description ... 59

electrical circuit can be described by the equation (5) or by the equation (6). This means that the descriptions (5) and (6) are independent on the observer. Each of them can be considered the description of the dynamic of current intensity in the considered RLC series circuit.

VON-SCHWEIDLER DESCRIPTION OF THE ELECTRON TRANSPORT IN ELECTRIC CIRCUIT USING INTEGER ORDER DERIVATIVE IS OBJECTIVE

In the description of the dynamic of the current intensity in the above considered RLC series electric circuit, the Curie effect is completly ignored. Namely, in [4], J. Curie shows that, if in case of a dielectric material at the moment of time $t=0$ a constant V_0 external voltage is applied, then the current intensity produced in that material is $i_C(t) = V_0/(h \cdot t^\alpha)$ with $0 < \alpha < 1$, $t > 0$ and h a constant related to the capacitance of the capacitor and the kind of dielectric.

Schweidler in [5] shows that, due to the Curie effect, the current intensity in the considered RLC series circuit, due to a constant external voltage V_0, is in fact $i(t) = I(t) + i_C(t)$, where $i_C(t) = V_0/(h \cdot t^\alpha)$ is the Curie current, $I = I(t)$ is the ohmic current and verifies the equation $L \cdot \dfrac{d^2 I}{dt^2} + R \cdot \dfrac{dI}{dt} + \dfrac{I}{C} = 0$. So, the differential equation governing the dynamic of the current intensity $i(t)$ in the circuit is:

$$L \cdot \frac{d^2 i}{dt^2} + R \cdot \frac{di}{dt} + \frac{i}{C} + \frac{V_0}{h} \cdot \left[-L \cdot \alpha \cdot (\alpha + 1) + R \cdot \alpha \cdot t - \frac{t^2}{C} \right] \cdot \frac{1}{t^{\alpha+2}} = 0 \qquad (8)$$

In other words, if the events "the external voltage is applied to the RLC circuit" and "the observer O chronometer starts" are simultaneous, then the current intensity variation, described by the observer O with the function $i = i(t_M)$, verifies the following equation:

$$L \cdot \frac{d^2 i}{dt_M^2} + R \cdot \frac{di}{dt_M} + \frac{i}{C} + \frac{V_0}{h} \cdot \left[-L \cdot \alpha \cdot (\alpha+1) + R \cdot \alpha \cdot t_M - \frac{t_M^2}{C} \right] \cdot \frac{1}{t_M^{\alpha+2}} = 0 \quad (9)$$

Note that this equation describes the dynamic if and only if the constant V_0 external voltage is applied at the moment M_O i.e., $t_{M_O} = 0$.

Observer $O*$, for which $t_{M_{O*}} > 0$, describes the electron transport in the same electric circuit with the equation

$$L \cdot \frac{d^2 i^*}{dt^*_M{}^2} + R \cdot \frac{di^*}{dt^*_M} + \frac{i^*}{C} +$$
$$\frac{V_0}{h} \cdot \left[-L \cdot \alpha \cdot (\alpha+1) + R \cdot \alpha \cdot t^*_M - \frac{t^*_M{}^2}{C} \right] \cdot \frac{1}{t^*_M{}^{\alpha+2}} = 0 \quad (10)$$

Equations (9) and (10) are different, but their solutions describe the current intensity variation in the same RLC circuit under the action of the same external source voltage. This can be proven showing that if $i = i(t_M)$ is a solution of (9), then $i^* = i^*(t^*_M)$, defined by (3), is a solution of (10) and if $i^* = i^*(t^*_M)$ is a solution of (10), then $i = i(t_M)$, defined by (4), is a solution of (9).

CAPUTO, RIEMANN-LIOUVILLE FRACTIONAL ORDER DERIVATIVE, DEFINED WITH INTEGRAL REPRESENTATION ON FINITE INTERVAL AND GENERAL LIOUVILLE-CAPUTO, GENERAL RIEMANN-LIOUVILLE

According to [6], [7] for a continuously differentiable function $f:[0,\infty) \to R$ the Caputo fractional derivative of order $\alpha > 0$, is defined with formula:

Objective and Nonobjective Mathematical Description ...

$$^{[0,t]}D_C^\alpha f(t) = \frac{1}{\Gamma(n-\alpha)} \cdot \int_0^t \frac{\frac{d^n f}{d\tau^n}(\tau)}{(t-\tau)^{\alpha+1-n}} d\tau \qquad (11)$$

Remark that the derivative defined with (11) was considered by other people before Caputo, like Gherasimov (see [8]). So, the name of Caputo, given in this chapter, maybe is not appropriate.

For a continuously differentiable function $f:[0,\infty) \to R$ the Riemann-Liouville fractional derivative of order $\alpha > 0$, according to [7], defined with formula:

$$^{[0,t]}D_{R-L}^\alpha f(t) = \frac{1}{\Gamma(n-\alpha)} \cdot \frac{d^n}{dt^n} \int_0^t \frac{f(\tau)}{(t-\tau)^{\alpha+1-n}} d\tau \qquad (12)$$

According to [8], the derivative having integral representation on infinite interval was proposed by Liouville in 1832. Some people call it general Liouville-Caputo derivative. Some people call it simply Liouville derivative ([8]).

For a continuously differentiable function $f:(-\infty,\infty) \to R$ the general Liouville - Caputo fractional order derivative of order $\alpha > 0$ is defined in [8] with formula:

$$^{(-\infty,t]}D_{L-C}^\alpha f(t) = \frac{1}{\Gamma(n-\alpha)} \cdot \int_{-\infty}^t \frac{\frac{d^n f}{d\tau^n}(\tau)}{(t-\tau)^{\alpha+1-n}} d\tau \qquad (13)$$

The general Riemann-Liouville fractional order derivative of order $\alpha > 0$ is defined in [8] with formula

$$^{(-\infty,t]}D_{R-L}^\alpha f(t) = \frac{1}{\Gamma(n-\alpha)} \cdot \frac{d^n}{dt^n} \int_{-\infty}^t \frac{f(\tau)}{(t-\tau)^{\alpha+1-n}} d\tau \qquad (14)$$

Here Γ is the Euler gamma function and $n = [\alpha]+1$, $[\alpha]$ being the integer part of α.

NONBJECTIVE DESCRIPTION OF THE ELECTRON TRANSPORT IN ELECTRIC CIRCUIT USING CAPUTO FRACTIONAL ORDER DERIVATIVE DEFINED WITH FORMULA (11) OR RIEMANN-LIOUVILLE FRACTIONAL ORDER DERIVATIVE DEFINED WITH FORMULA (12)

In [9] the authors present a classical derivation of fractional order circuits models. Generalized constitutive equations in terms of fractional order Riemann-Liouville derivatives are introduced in the Maxwell's equations. The Kirchhoff voltage law is applied in a RLC configuration and a fractional order differential equation is obtained for RLC circuit with Caputo derivatives. The authors consider that the Curie effect is a particular case of the transport phenomena in the complex media presented in [9].

In this section we show that the description of electron transport in a RLC series circuit by substituting in (5) and (6) the integer order derivatives of the current intensity with Caputo or Riemann-Liouville fractional order derivatives and the constants R, L, C with other constants depending on the order of derivatives, is nonobjective.

Consider first the case $1 < \alpha < 2$, $0 < \beta < 1$ and the substitution of the integer order derivatives of the current intensity with Caputo fractional order derivative.

After the substitution, for the observers O and O^* equations (5) and (6) become:

$$L_\alpha \cdot {}^{[0,t_M]}D_C^\alpha i(t_M) + R_\beta \cdot {}^{[0,t_M]}D_C^\beta i(t_M) + \frac{1}{C_{\alpha\beta}} \cdot i(t_M) = \frac{dV}{dt_M} \quad (15)$$

$$L_\alpha \cdot {}^{[0,t^*_M]}D_C^\alpha i^*(t^*_M) + R_\beta \cdot {}^{[0,t^*_M]}D_C^\beta i^*(t^*_M) + \frac{1}{C_{\alpha\beta}} \cdot i^*(t^*_M) = \frac{dV^*}{dt^*_M} \quad (16)$$

For $t_M > t_{M_O^*} > 0$ equalities:

$$^{[0,t_M]}D_C^\alpha i(t_M) = {}^{[0,t^*_M]}D_C^\alpha i*(t^*_M) + \frac{1}{\Gamma(2-\alpha)} \cdot \int_0^{t_{M_O^*}} \frac{\frac{d^2 i}{d\tau^2}(\tau)}{(t_M - \tau)^{\alpha-1}} d\tau$$

$$^{[0,t_M]}D_C^\beta i(t_M) = {}^{[0,t^*_M]}D_C^\beta i*(t^*_M) + \frac{1}{\Gamma(1-\beta)} \cdot \int_0^{t_{M_O^*}} \frac{\frac{di}{d\tau}(\tau)}{(t_M - \tau)^{\beta}} d\tau$$

imply that objectivity holds if and only if the following condition is verified:

$$\frac{L_\alpha}{\Gamma(2-\alpha)} \cdot \int_0^{t_{M_O^*}} \frac{\frac{d^2 i}{d\tau^2}(\tau)}{(t_M - \tau)^{\alpha-1}} d\tau + \frac{R_\beta}{\Gamma(1-\beta)} \cdot \int_0^{t_{M_O^*}} \frac{\frac{di}{d\tau}(\tau)}{(t_M - \tau)^{\beta}} d\tau = 0 \quad (17)$$

Condition (17) in general is not verified. So, this kind of electron transport description is nonobjective.

In the case $1 < \alpha < 2$, $0 < \beta < 1$, the substitution of the integer order derivatives of the current intensity with Riemann-Liouville fractional order derivatives leads to the following objectivity condition:

$$\frac{L_\alpha}{\Gamma(2-\alpha)} \cdot \frac{d^2}{dt_M^2} \int_0^{t_{M_O^*}} \frac{i(\tau)}{(t_M - \tau)^{\alpha-1}} d\tau + \frac{R_\beta}{\Gamma(1-\beta)} \cdot \frac{d}{dt_M} \int_0^{t_{M_O^*}} \frac{i(\tau)}{(t_M - \tau)^{\beta}} d\tau = 0 \quad (18)$$

Condition (18) in general is not verified. So, this kind of description is nonobjective.

OBJECTIVE DESCRIPTION OF THE ELECTRON TRANSPORT IN ELECTRIC CIRCUIT USING GENERAL LIOUVILLE–CAPUTO FRACTIONAL ORDER DERIVATIVE DEFINED WITH FORMULA (13) OR GENERAL ROEMANN-LIOUVILLE FRACTIONAL ORDER DERIVATIVES DEFINED WITH FORMULA (14)

Consider first the case $1 < \alpha < 2$, $0 < \beta < 1$ and the substitution of the integer order derivatives of the current intensity with general Liouville-Caputo fractional order derivatives.

After the substitution for the observers O and O^* equations (5) and (6) become:

$$L_\alpha \cdot {}^{(-\infty,t_M]}D_{L-C}^\alpha i(t_M) + R_\beta \cdot {}^{(-\infty,t_M]}D_{L-C}^\beta i(t_M) + \frac{1}{C_{\alpha\beta}} \cdot i(t_M) = \frac{dV}{dt_M} \quad (19)$$

$$L_\alpha \cdot {}^{(-\infty,t^*_M]}D_{L-C}^\alpha i^*(t^*_M) + R_\beta \cdot {}^{(-\infty,t^*_M]}D_{L-C}^\beta i^*(t_M) + \frac{1}{C_{\alpha\beta}} \cdot i^*(t^*_M) = \frac{dV^*}{dt^*_M} \quad (20)$$

Equalities:

$${}^{(-\infty,t_M]}D_{L-C}^\alpha i(t_M) = {}^{(-\infty,t^*_M]}D_{L-C}^\alpha i^*(t^*_M)$$

$${}^{(-\infty,t_M]}D_{L-C}^\beta i(t_M) = {}^{(-\infty,t^*_M]}D_{L-C}^\beta i^*(t^*_M)$$

imply that objectivity holds.

In the case $1 < \alpha < 2$, $0 < \beta < 1$ and the substitution of the integer order derivatives of the current intensity with general Riemann-Liouville fractional order derivatives, the equalities:

$$^{(-\infty,t_M]}D_{R-L}{}^{\alpha}i(t_M) = {}^{(-\infty,t^*_M]}D_{R-L}{}^{\alpha}i*(t^*_M)$$

$$^{(-\infty,t_M]}D_{R-L}{}^{\beta}i(t_M) = {}^{(-\infty,t^*_M]}D_{R-L}{}^{\beta}i*(t^*_M)$$

imply that objectivity holds.

OBJECTIVE DESCRIPTION OF THE ION TRANSPORT ACROSS A PASSIVE MEMBRANE OF A BIOLOGICAL NEURON USING INTEGER ORDER DERIVATIVE

In [10] the ion transport across a passive membrane of a biological neuron is presented as the electron transport in a parallel resistor-capacitor circuit, assuming ideal capacitive behavior. The transport is described by a real valued and real variable function V_m, which represents the membrane voltage variation due to the applied stimulus current.

In terms of observer O the membrane voltage variation $V_m = V_m(t_M)$ verifies the differential equation

$$C_m \cdot \frac{dV_m}{dt_m} + \frac{V_m}{R_m} = I(t_M)_{appl} \qquad (21)$$

where: C_m is the membrane capacitance, R_m is the membrane resistance, $I(t_M)_{appl}$ represents the applied stimulus current intensity.

In terms of observer $O*$ the membrane voltage variation $V*_m = V*_m(t*_M)$ verifies the differential equation

$$C_m \cdot \frac{d*V_m}{d*t_m} + \frac{V*_m}{R_m} = I*(t*_M)_{appl} \qquad (22)$$

here: C_m is the membrane capacitance, R_m is the membrane resistance, $I*(t*_M)_{appl}$ represents the applied stimulus current intensity.

Since the applied stimulus current intensity is independent on observer, the applied stimulus intensity verifies:

$$I*(t*_M)_{appl} = I(t*_M + t_{M_{O^*}})_{appl}. \qquad (23)$$

Equations (21) and (23) are different, but their solutions describe the voltage variation in the same R-C parallel circuit under the action of the same applied stimulus current intensity. This can be proven showing that if $V_m = V_m(t_M)$ is a solution of (21), then $V_m* = V_m*(t*_M)$, defined by $V_m*(t*_M) = V_m(t*_M + t_{M_{O^*}})$, is a solution of (22) and if $V_m* = V_m*(t*_M)$ is a solution of (22), then $V_m = V_m(t_M)$, defined by $V_m(t_M) = V*_m(t_M + t*_{M_O})$, is a solution of (21).

NONOBJECTIVE DESCRIPTION OF THE ION TRANSPORT ACROSS A PASSIVE MEMBRANE OF A BIOLOGICAL NEURON CELL USING CAPUTO FRACTIONAL ORDER DERIVATIVE DEFINED WITH FORMULA (11) OR RIEMANN-LIOUVILLE FRACTIONAL ORDER DERIVATIVE DEFINED WITH FORMULA (12)

In [10], the author undertakes a study of the case when the description of the ion transport through a passive membrane verifies a fractional order differential equation. The author of [10] underlines that: "the physiological source of such non-ideal capacitive behavior is not known, but this may arise due to heterogeneities in the dielectric properties of the membrane".

In the case of Caputo fractional order derivative, defined with formula (11), the neuron membrane voltage V_m verifies the fractional order differential equation:

$$C_\alpha \cdot {}^{[0,t_M]}D_C^\alpha V_m(t_M) + V_m(t_M)/R_a = I(t_M)_{appl} \qquad (24)$$

It is easy to show that the description of the ion transport across the passive membrane of neuron with equation (24) is objective if and only if the next equality holds:

$$\frac{1}{\Gamma(1-\alpha)} \int_0^{t_{MQ^*}} \frac{\frac{dV_m}{d\tau}(\tau)}{(t_M - \tau)^\alpha} d\tau = 0 \qquad (25)$$

In general, condition (25) is not fullfilled. So, the description is not objective.

In the case of the use of Riemann-Liouville fractional order derivative, defined with formula (12), the membrane voltage V_m verifies the equation:

$$C_\alpha \cdot {}^{[0,t_M]}D_{R-L}^\alpha V_m(t_M) + V_m(t_M)/R_a = I(t_M)_{appl} \qquad (26)$$

The description of the ion transport across the passive membrane of the neuron with equations (26) is objective if and only if the next equality holds:

$$\frac{1}{\Gamma(1-\alpha)} \cdot \frac{d}{dt} \int_0^{t_{MQ^*}} \frac{V_m(\tau)}{(t_M - \tau)^\alpha} d\tau = 0 \qquad (27)$$

In general, equality (27) is not fullfilled. So, the description is nonobjective.

OBJECTIVE DESCRIPTION OF THE ION TRANSPORT ACROSS A PASSIVE MEMBRANE OF A BIOLOGICAL NEURON USING GENERAL LIOUVILLE-CAPUTO FRACTIONAL ORDER DERIVATIVE DEFINED WITH FORMULA (13) OR GENERAL RIEMANN-LIOUVILLE FRACTIONAL ORDER DERIVATIVE DEFINED WITH FORMULA (14)

In the case of general Liouville-Caputo fractional order derivative, defined with formula (13), the membrane voltage V_m verifies the fractional order differential equation:

$$C_a \cdot {}^{(-\infty, t_M]}D_{L-C}{}^\alpha V_m(t_M) + V_m(t_M)/R_a = I(t_M)_{appl} \qquad (28)$$

It is easy to show that the description of the ion transport through the passive membrane with equation (28) is objective, because the next equality holds:

$$^{(-\infty, t_M]}D_{L-C}{}^\alpha V_m(t_M) = {}^{(-\infty, t^*_M]}D_{L-C}{}^\alpha V_m^*(t^*_M) \qquad (29)$$

In the case of the use of Riemann-Liouville fractional order derivative, defined with formula (14), the membrane voltage V_m verifies the equation:

$$C_a \cdot {}^{(-\infty, t_M]}D_{R-L}{}^\alpha V_m(t_M) + V_m(t_M)/R_a = I(t_M)_{appl} \qquad (30)$$

Description of the ion transport across the passive membrane of neuron with equations (30) is objective, because the next equality holds:

$$^{(-\infty, t_M]}D_{R-L}{}^\alpha V_m(t_M) = {}^{(-\infty, t^*_M]}D_{R-L}{}^\alpha V_m^*(t^*_M) \qquad (31)$$

HODGKIN-HUXLEY DESCRIPTION OF THE ION TRANSPORT ACROSS THE MEMBRANE OF THE SQUID GIANT AXON WITH INTEGER ORDER DERIVATIVE IS OBJECTIVE

In Hodgkin–Huxley description [11] the biophysical characteristic of the squid giant axon membrane appear as elements of an electric circuit. The lipid bilayer membrane is represented as a capacitance (C_m). Voltage-gated and leak ion channels are represented by nonlinear (g_n) and linear (g_L) conductances, respectively. The electrochemical gradients driving the flow of ions are represented by batteries (E), and ion pumps and exchangers are represented by current sources (I_p).

The sodium and potassium ions transport across the membrane of the squid giant axon is described in terms of observer O with a set of four real valued and real variables functions: $V_m = V_m(t_M)$, $m = m(t_M)$, $h = h(t_M)$, and $n = n(t_M)$. Here: V_m represents the membrane potential; m represents the sodium current activation gating variable; h represents the sodium current inactivation gating variable; n represents the potassium current activation gating variable.

These functions check the following differential equations:

$$C_m \cdot dV_m / dt_M + \overline{g}_{Na} \cdot m^3 \cdot h \cdot (V_m - V_{Na}) + \overline{g}_K \cdot n^4 \cdot (V_m - V_K) +$$
$$\overline{g}_L \cdot (V_m - V_L) = I_{appl}(t_M)$$
$$dn / dt_M = \alpha_n(V_m) \cdot (1-n) - \beta_n(V_m) \cdot n$$
$$dm / dt_M = \alpha_m(V_m) \cdot (1-m) - \beta_m(V_m) \cdot m \quad (32)$$
$$dh / dt_M = \alpha_h(V_m) \cdot (1-h) - \beta_h(V_m) \cdot h$$

where C_m, $\overline{g}_{Na}$, $\overline{g}_K$, $\overline{g}_L$, V_{Na}, V_K, V_L are constants.

In the original paper of Hudgkin-Huxley [11] the functions α and β are given by:

$$\alpha_n(V_m) = \frac{0.01(10 - V_m)}{\exp\left(\frac{10 - V_m}{10}\right) - 1} \quad \alpha_m(V_m) = \frac{0.1(25 - V_m)}{\exp\left(\frac{25 - V_m}{10}\right) - 1} \quad \alpha_h(V_m) = 0.07 \exp\left(\frac{-V_m}{20}\right)$$
$$\beta_n(V_m) = 0.125 \exp\left(\frac{-V_m}{80}\right) \quad \beta_m(V_m) = 4 \exp\left(\frac{-V_m}{18}\right) \quad \beta_h(V_m) = \frac{1}{\exp\left(\frac{30 - V_m}{10}\right) + 1} \quad (33)$$

In terms of observer O^* the sodium and potassium ions transport across the squid giant axon membrane is described by the functions $V^*_m = V^*_m(t^*_M), m^* = m^*(t^*_M), h^* = h^*(t^*_M), n^* = n^*(t^*_M)$ and check the differential equations:

$$C_m \cdot dV^*_m / dt^*_M + \bar{g}_{Na} \cdot m^{*3} \cdot h^* \cdot (V^*_m - V_{Na}) + \bar{g}_K \cdot n^{*4} \cdot (V^*_m - V_K) + \bar{g}_L \cdot (V^*_m - V_L) = I_{appl}^*(t^*_M)$$
$$dn^*/dt^*_M = \alpha_n(V^*_m) \cdot (1 - n^*) - \beta_n(V^*_m) \cdot n^*$$
$$dm^*/dt^*_M = \alpha_m(V^*_m) \cdot (1 - m^*) - \beta_m(V^*_m) \cdot m^* \qquad (34)$$
$$dh^*/dt^*_M = \alpha_h(V^*_m) \cdot (1 - h^*) - \beta_h(V^*_m) \cdot h^*$$

Since the applied stimulus current intensity is independent on observer, the intensity verifies

$$I^*(t^*_M)_{appl} = I(t^*_M + t_{M_{O^*}})_{appl}. \qquad (35)$$

Equations (32) and (34) are different, but their solutions describe the voltage variations in the same squid giant axon membrane under the action of the same applied stimulus current intensity. This can be proven showing that: if $V_m = V_m(t_M), m = m(t_M), h = h(t_M), n = n(t_M)$ is a solution of (32), then $V_m^* = V_m^*(t^*_M)$, $m^* = m^*(t^*_M), h^* = h^*(t^*_M), n^* = n^*(t^*_M)$, defined by: $V_m^*(t^*_M) = V_m(t^*_M + t_{M_{O^*}})$, $m^*(t^*_M) = m(t^*_M + t_{M_{O^*}}), h^*(t^*_M) = h(t^*_M + t_{M_{O^*}}), n^*(t^*_M) = n(t^*_M + t_{M_{O^*}})$, is a solution of (34) and, if $V_m^* = V_m^*(t^*_M), m^* = m^*(t^*_M), h^* = h^*(t^*_M)$, $n^* = n^*(t^*_M)$ is a solution of (34), then $V_m = V_m(t_M), m = m(t_M), h = h(t_M)$, $n = n(t_M)$, defined by $V_m(t_M) = V^*_m(t_M + t^*_{M_O}), m(t_M) = m^*(t_M + t^*_{M_O})$, $h(t_M) = h^*(t_M + t^*_{M_O}), n(t_M) = n^*(t_M + t_{M_O})$, is a solution of (32). So, the description is objective.

HODGKIN-HUXLEY TYPE DESCRIPTION OF THE ION TRANSPORT ACROSS THE MEMBRANE OF THE SQUID GIANT AXON USING CAPUTO OR RIEMANN-LIOUVILLE FRACTIONAL ORDER DERIVATIVES DEFINED WITH INTEGRAL REPRESENTATION ON A FINITE INYERVAL, IS NONOBJECTIVE

In [10], the author undertakes a study using for the description of the ion transport across the squid giant axon membrane Caputo or Riemann-Liouville fractional order derivative, defined with integral representation on a finite interval. This type of description in [10] is called the fractional order membrane patch model.

In the case of Caputo fractional order derivatives, defined with formula (11) and observer O, instead of equations (32) in the fractional order membrane patch model the following equations are used:

$$C_m^\alpha \cdot {}^{[0,t_M]}D_C^\alpha V_m(t_M) + \bar{g}_{Na} \cdot m^3 \cdot h \cdot (V_m - V_{Na}) + \bar{g}_K \cdot n^4 \cdot (V_m - V_K) +$$
$$\bar{g}_L \cdot (V_m - V_L) = I_{appl}(t_M)$$
$$dn/dt_M = \alpha_n(V_m) \cdot (1-n) - \beta_n(V_m) \cdot n$$
$$dm/dt_M = \alpha_m(V_m) \cdot (1-m) - \beta_m(V_m) \cdot m \qquad (36)$$
$$dh/dt_M = \alpha_h(V_m) \cdot (1-h) - \beta_h(V_m) \cdot h$$

It is easy to see that the transport description with equations (36) is objective if and only if the following equality holds:

$$\frac{1}{\Gamma(1-\alpha)} \int_0^{t_{MO^*}} \frac{\frac{dV_m}{d\tau}(\tau)}{(t_M - \tau)^\alpha} d\tau = 0 \qquad (37)$$

In general, condition (37) is not fullfilled. So, the fractional-order membrane patch model, which uses Caputo fractional order derivative, defined with formula (11), is nonobjective.

In the case of Riemann-Liouville fractional order derivatives, defined with formula (12) and observer O, in the fractional order membrane patch model, instead of equations (32), the following equations are used:

$$C_m^\alpha \cdot {}^{[0,t_M]}D_{R-L}^\alpha V_m(t_M) + \bar{g}_{Na} \cdot m^3 \cdot h \cdot (V_m - V_{Na}) + \bar{g}_K \cdot n^4 \cdot (V_m - V_K) +$$
$$\bar{g}_L \cdot (V_m - V_L) = I_{appl}(t_M)$$
$$dn/dt_M = \alpha_n(V_m) \cdot (1-n) - \beta_n(V_m) \cdot n$$
$$dm/dt_M = \alpha_m(V_m) \cdot (1-m) - \beta_m(V_m) \cdot m \qquad (38)$$
$$dh/dt_M = \alpha_h(V_m) \cdot (1-h) - \beta_h(V_m) \cdot h$$

It is easy to see that the transport description with equations (38) is objective if and only if the following equality holds:

$$\frac{1}{\Gamma(1-\alpha)} \frac{d}{dt_M} \int_0^{t_{MO^*}} \frac{V_m(\tau)}{(t_M - \tau)^\alpha} d\tau = 0 \qquad (39)$$

In general, condition (39) is not fullfilled. So, the fractional-order membrane patch model which uses Riemann-Liouville fractional order derivative, defined with formula (12), is nonobjective.

HODGKIN-HUXLEY TYPE DESCRIPTION OF THE ION TRANSPORT ACROSS THE MEMBRANE OF THE SQUID GIANT AXON USING GENERAL LIOUVILLE-CAPUTO OR GENERAL RIEMANN-LIOUVILLE FRACTIONAL ORDER DERIVATIVE, DEFINED WITH INTEGRAL REPRESENTATION ON INFINITE INTERVAL, IS OBJECTIVE

If in the fractional-order membrane patch model, instead of Caputo fractional order derivative, general Liouville-Caputo fractional order derivative (formula

(13)) is used, then in case of observer O the transport is described by the equations:

$$C_m^\alpha \cdot {}^{(-\infty,t_M]}D_{L-C}^\alpha V_m(t_M) + \bar{g}_{Na} \cdot m^3 \cdot h \cdot (V_m - V_{Na}) + \bar{g}_K \cdot n^4 \cdot (V_m - V_K) +$$
$$\bar{g}_L \cdot (V_m - V_L) = I_{appl}(t_M)$$
$$dn/dt_M = \alpha_n(V_m) \cdot (1-n) - \beta_n(V_m) \cdot n$$
$$dm/dt_m = \alpha_m(V_m) \cdot (1-m) - \beta_m(V_m) \cdot m \qquad (40)$$
$$dh/dt_m = \alpha_h(V_m) \cdot (1-h) - \beta_h(V_m) \cdot h$$

The transport description with equations (40) is objective, because the following equality holds:

$${}^{(-\infty,t_M]}D_{L-C}^\alpha V_m(t_M) = {}^{(-\infty,t^*_M]}D_{L-C}^\alpha V^*_m(t^*_M) \qquad (41)$$

If in the fractional-order membrane patch model, instead of Riemann-Liouville fractional order derivative, general Riemann-Liouville fractional order derivative (formula (14)) is used, then in case of observer O the transport is described by the equations:

$$C_m^\alpha \cdot {}^{(-\infty,t_M]}D_{R-L}^\alpha V_m(t_M) + \bar{g}_{Na} \cdot m^3 \cdot h \cdot (V_m - V_{Na}) + \bar{g}_K \cdot n^4 \cdot (V_m - V_K) +$$
$$\bar{g}_L \cdot (V_m - V_L) = I_{appl}(t_M)$$
$$dn/dt_M = \alpha_n(V_m) \cdot (1-n) - \beta_n(V_m) \cdot n$$
$$dm/dt_M = \alpha_m(V_m) \cdot (1-m) - \beta_m(V_m) \cdot m \qquad (42)$$
$$dh/dt_M = \alpha_h(V_m) \cdot (1-h) - \beta_h(V_m) \cdot h$$

The transport description with equations (42) is objective, because the following equality holds:

$$^{(-\infty,t_M]}D_{R-L}{}^\alpha V_m(t_M) = {}^{(-\infty,t^*_M]}D_{R-L}{}^\alpha V^*_m(t^*_M) \qquad (43)$$

MORRIS-LECAR DESCRIPTION OF THE ION TRANSPORT ACROSS THE MEMBRANE OF THE BARNACLE GIANT MUSCLE FIBER WITH INTEGER ORDER DERIVATIVE IS OBJECTIVE

Morris and Lecar in [12] show that voltage-clamp studies of the barnacle muscle made in [13]- [17] indicate that the fiber possesses a simply conductance system consisting of voltage dependent Ca^{++} and K^+ channels, neither of which innactivates appreciably. Current clamp studies [16] and [18], however, show complicated oscillatory voltage behavior. The mathematical study developed in [12] reveals that this simple system can predict much of the brancle fiber behavior, although the simplest model fails to explain some areas of behavior.

The system of differential equations describing the votage oscillations, considered in [12], is the following:

$$\begin{aligned} C \cdot dV/dt &= I(t) - g_{Ca} \cdot M \cdot [V - V_{Ca}] - g_K \cdot N \cdot [V - V_K] - g_L \cdot [V - V_L] \\ dM/dt &= \lambda_M(V) \cdot [M_\infty(V) - M] \\ dN/dt &= \lambda_N(V) \cdot [N_\infty(V) - N] \end{aligned} \qquad (44)$$

The meaning of the symbols appearing in the above systems can be found in [12].

It is easy to see that the description of the voltage oscillation by equations (44) is objective.

MORRIS-LECAR TYPE DESCRIPTION OF THE ION TRANSPORT ACROSS THE MEMBRANE OF THE BARNACLE GIANT MUSCLE FIBER USING CAPUTO OR RIEMANN-LIOUVILLE FRACTIONAL ORDER DERIVATIVE, DEFINED ON FINITE INTERVAL, IS NONOBJECTIVE

In the case of Caputo fractional order derivatives defined on finite interval, in terms of observer O, equations (44) lead to the equations:

$$C_\alpha \cdot {}^{[0,t_M]}D_C^\alpha V(t_M) = I(t_M) - g_{Ca} \cdot M \cdot [V - V_{Ca}] - g_K \cdot N \cdot [V - V_K] - g_L \cdot [V - V_L]$$
$$[0,t_M]D_C^\beta M(t_M) = \lambda_M(V) \cdot [M_\infty(V) - M] \qquad (45)$$
$$[0,t_M]D_C^\gamma N(t_M) = \lambda_N(V) \cdot [N_\infty(V) - N]$$

It is easy to shown that the description of the voltage oscillation by equations (45) is objective if and only if the following conditions are verified:

$$\frac{C_\alpha}{\Gamma(1-\alpha)} \cdot \int_0^{t_{MO^*}} \frac{V'(\tau)}{(t_M - \tau)^\alpha} d\tau = 0; \quad \frac{1}{\Gamma(1-\beta)} \cdot \int_0^{t_{MO^*}} \frac{N'(\tau)}{(t - \tau)^\beta} d\tau = 0;$$
$$\frac{1}{\Gamma(1-\gamma)} \cdot \int_0^{t_{MO^*}} \frac{M'(\tau)}{(t - \tau)^\gamma} d\tau = 0 \qquad (46)$$

In general, the above conditions are not fullfilled. Therefore, the Morris – Lecar model considered in [19] is nonobjective.

In the case of Riemann-Liouville fractional order derivatives, defined on finite interval, the use of the description of the voltage oscillation is objective if and only if the following conditions are verified:

$$\frac{C_\alpha}{\Gamma(1-\alpha)} \cdot \frac{d}{dt_M} \int_0^{t_{MO^*}} \frac{V(\tau)}{(t_M - \tau)^\alpha} d\tau = 0 \quad \frac{1}{\Gamma(1-\beta)} \cdot \frac{d}{dt_M} \int_0^{t_{MO^*}} \frac{N(\tau)}{(t_M - \tau)^\beta} d\tau = 0$$

$$\frac{1}{\Gamma(1-\gamma)} \cdot \frac{d}{dt_M} \int_0^{t_{MO^*}} \frac{M(\tau)}{(t_M-\tau)^\gamma} d\tau = 0 \qquad (47)$$

In general, conditions (47) are not fullfilled. Therefore, the Morris – Lecar model which uses Riemann-Liouville fractional order derivatives, defined on finite interval, is nonobjective.

MORRIS-LECAR TYPE DESCRIPTION OF THE ION TRANSPORT ACROSS THE MEMBRANE OF THE BARNACLE GIANT MUSCLE FIBER USING GENERAL LIOUVILLE-CAPUTO OR GENERAL RIEMANN-LIOUVILLE FRACTIONAL ORDER DERIVATIVES, DEFINED ON INFINITE INTERVAL, IS OBJECTIVE

In the case of general Liouville-Caputo fractional order derivatives defined on infinite interval, in terms of observer O equations (44), lead to the equations:

$$C_\alpha \cdot {}^{(-\infty,t_M)}D_{L-C}^\alpha V(t_M) = I(t_M) - g_{Ca} \cdot M \cdot [V-V_{Ca}] - g_K \cdot N \cdot [V-V_K] - g_L \cdot [V-V_L]$$

$$ {}^{(-\infty,t_M]}D_{L-C}^\beta M(t_M) = \lambda_M(V) \cdot [M_\infty(V) - M]$$

$$C_\alpha \cdot {}^{(-\infty,t_M)}D_{L-C}^\alpha V(t_M) = I(t_M) - g_{Ca} \cdot M \cdot [V-V_{Ca}] - g_K \cdot N \cdot [V-V_K] - g_L \cdot [V-V_L]$$

$$ {}^{(-\infty,t_M]}D_{L-C}^\beta M(t_M) = \lambda_M(V) \cdot [M_\infty(V) - M] \qquad (48)$$

$$ {}^{[(-\infty,t_M]}D_{L-C}^\gamma N(t_M) = \lambda_N(V) \cdot [N_\infty(V) - N]$$

The description of the voltage oscillation by equations (48) is objective, because the following equalities hold:

$$ {}^{(-\infty,t_M]}D_{L-C}^\alpha V_m(t_M) = {}^{(-\infty,t^*_M]}D_{L-C}^\alpha V^*_m(t^*_M)$$

$$^{(-\infty,t_M]}D_{L-C}{}^{\beta}V_m(t_M) = {}^{(-\infty,t^*_M]}D_{L-C}{}^{\beta}V^*_m(t^*_M) \qquad (49)$$
$$^{(-\infty,t_M]}D_{L-C}{}^{\gamma}V_m(t_M) = {}^{(-\infty,t^*_M]}D_{L-C}{}^{\gamma}V^*_m(t^*_M)$$

In the case of general Riemann-Liouville fractional order derivatives, defined on infinite interval, in terms of observer O, equations (44) lead to the equations:

$$C_\alpha \cdot {}^{(-\infty,t_M)}D_{R-L}{}^{\alpha}V(t_M) = I(t_M) - g_{Ca} \cdot M \cdot [V - V_{Ca}] - g_K \cdot N \cdot [V - V_K] -$$
$$g_L \cdot [V - V_L]$$
$$^{(-\infty,t_M]}D_{R-L}{}^{\beta}M(t_M) = \lambda_M(V) \cdot [M_\infty(V) - M] \qquad (50)$$
$$^{[(-\alpha t_M]}D_{R-L}{}^{\gamma}N(t_M) = \lambda_N(V) \cdot [N_\infty(V) - N]$$

The description of the voltage oscillation by equations (48) is objective, because the following equalities hold:

$$^{(-\infty,t_M]}D_{R-L}{}^{\alpha}V_m(t_M) = {}^{(-\infty,t^*_M]}D_{R-L}{}^{\alpha}V^*_m(t^*_M) \qquad (51)$$
$$^{(-\infty,t_M]}D_{R-L}{}^{\beta}V_m(t_M) = {}^{(-\infty,t^*_M]}D_{R-L}{}^{\beta}V^*_m(t^*_M)$$
$$^{(-\infty,t_M]}D_{R-L}{}^{\gamma}V_m(t_M) = {}^{(-\infty,t^*_M]}D_{R-L}{}^{\gamma}V^*_m(t^*_M)$$

OBJECTIVE DESCRIPTION OF THE ION TRANSPORT ALONG DENDRITE AND AXON HAVING PASSIVE MEMBRANE IN CLASSICAL CABEL THEORY (AXON MODEL) USING INTEGER ORDER DERIVATIVES

In classical mathematical cable theory (axon model see [10]) the neuron dendrite and axon are represented as segments of a line in the affine euclidian space E_1. For the description of the ion transport along passive dendrites and axons that receive synaptic inputs at different sites and times, observer O chooses a fixed point O on the line representing dendrite or axon, a reference frame $R_O = (O; \vec{e}_1)$, a moment of time M_O for fixing the origin of the time measurement and a unit [second] for time measuring. The coordinate of a point P on the line (representing the dendrite or axon), with respect to the reference frame R_O, is denoted by X_1. A moment of time M, which is earlier than M_O is represented by a negative real number $t_M < 0$, a moment of time M, which is later than M_O, is represented by a positive real number $t_M > 0$ and the moment of time M_O is represented by the real number $t_{M_O} = 0$. Observer O describes the ion transport along the passive dendrite and axon that receive synaptic inputs at different sites and times by a two variable real valued function $V_m = V_m(X_1, t_M)$ representing the voltage variation along the membrane. Here X_1 is the spatial coordinate with respect to the reference frame $R_O = (O; \vec{e}_1)$ and t_M represents the moment of time M for observer O.

Observer O^* chooses a fixed point O^* on the line representing dendrite or axon, a reference frame $R_{O^*} = (O^*; \vec{e}_1)$, a moment of time M_{O^*} for fixing the origin of the time measurement and a unit [second] for time measuring. The coordinate of the same point P on the line, representing the dendrite or axon, with respect to the reference frame R_{O^*}, is denoted by X^*_1. A moment of time M, which is earlier than M_{O^*}, is represented by a negative real number $t^*_M < 0$, a moment of time, M which is later than M_{O^*}, is represented by a positive real number $t^*_M > 0$ and the moment of time M_{O^*} is represented by the real number $t^*_{M_{O^*}} = 0$. Observer O^* describes the ion transport along passive dendrites and axons that receive synaptic inputs at different sites and times by a two variable real valued function $V^*_m = V^*_m(X^*_1, t^*_M)$, representing the voltage variation along the membrane. Here X^*_1 is the spatial coordinate with respect to the reference frame $R_{O^*} = (O^*; \vec{e}_1)$ and t^*_M represents the moment of time M for observer O^*.

Objective and Nonobjective Mathematical Description ...

Because the numbers X_1 and X^*_1 are the coordinates of the same point P with respect to R_O and R_{O^*}, respectively, for them the following relations hold:

$$X_1 = X^*_1 + X_{1O^*}; \quad X^*_1 = X_1 + X^*_{1O} \qquad (52)$$

here X_{1O^*} is the coordinate of O^* with respect to R_O and X^*_{1O} is the coordinate of O with respect to R_{O^*}.

Because the numbers t_M and t^*_M represent the same moment of time M for them the following relations hold:

$$t_M = t^*_M + t_{M_O^*}, \quad t^*_M = t_M + t^*_{M_O}. \qquad (53)$$

here $t_{M_O^*}$ is the number corresponding to the moment of time M_{O^*} in the system of observer O and $t^*_{M_O}$ is the number corresponding to the moment of time M_O in the system of observer O^*.

In terms of observer O the equation governing the membrane voltage evolution is:

$$c_m \cdot \frac{\partial V_m}{\partial t_M} + \frac{V_m}{r_m} - \frac{1}{r_l} \cdot \frac{\partial^2 V_m}{\partial X_1^2} = I(X_1, t_M)_{appl} \qquad (54)$$

here: c_m, r_m, r_l are constants representing capacitance, resistance and intracellular resistance, respectively (see [10]).

In terms of observer O^* the equation governing the membrane voltage evolution is:

$$c_m \cdot \frac{\partial V^*_m}{\partial t^*_M} + \frac{V^*_m}{r_m} - \frac{1}{r_l} \cdot \frac{\partial^2 V^*_m}{\partial X^{*2}_1} = I^*(X^*_1, t^*_M)_{appl} \qquad (55)$$

Since the applied stimulus current intensity is independent on observer the intensity verifies

$$I*(X*_1, t*_M)_{appl} = I(X*_1 + X_{1O*}, t*_M + t_{M_{O*}})_{appl}. \qquad (56)$$

Equations (54) and (55) are different, but their solutions have to describe the same voltage variations under the action of the same applied stimulus current intensity. This can be proven showing that if $V_m = V_m(X_1, t_M)$ is a solution of (54), then $V_m^* = V_m^*(X_1^*, t_M^*)$, defined $V_m^*(X_1^*, t_M^*) = V_m(X*_1 + X_{1O*}, t*_M + t_{M_{O*}})$ is a solution of (55) and if $V_m^* = V_m^*(X_1^*, t_M^*)$ is a solution of (55), then $V_m = V_m(X_1, t_M)$, defined by $V_m(X_1, t_M) = V_m^*(X_1 + X_{1O}^*, t_M + t_{M_O}^*)$, is a solution of (52). So, the description is objective.

DESCRIPTION OF THE ION TRANSPORT IN THE FRAMEWORK OF FRACTIONAL ORDER NERVE AXON MODEL USING CAPUTO OR RIEMANN-LIOUVILLE FRACTIONAL ORDER PARTIAL DERIVATIVE, DEFINED WITH INTEGRAL REPRESENTATION ON A FINITE INTERVAL, IS NONOBJECTIVE

In [10], the author undertakes a study, using for the description of the ion transport along the neuron axon partial Caputo or partial Riemann-Liouville fractional order derivatives, defined with integral representation on a finite interval. This type of description in [10] is called the fractional order nerve axon model.

In the case of Caputo fractional order derivatives and observer O, instead of equation (54) the following equation is used:

$$c_m^\alpha \cdot \frac{\partial^\alpha V_m}{\partial t_M^\alpha} + \frac{V_m}{r_m} - \frac{1}{r_i} \cdot \frac{\partial^2 V_m}{\partial X_1^2} = I(X_1, t_M)_{appl} \qquad (56)$$

where: $0 < \alpha < 1$ and

$$\frac{\partial^\alpha V_m}{\partial t_M^\alpha}(X_1, t_M) = {}^{[0,t_M]}D^\alpha_{C,t_M} V_m(X_1, t_M) = \frac{1}{\Gamma(1-\alpha)} \cdot \int_0^{t_M} \frac{\frac{\partial V_m}{\partial \tau}(X_1, \tau)}{(t_M - \tau)^\alpha} d\tau \qquad (57)$$

It is easy to shown that the transport description with equations (56) is objective if and only if the following equality holds:

$$\frac{1}{\Gamma(1-\alpha)} \int_0^{t_{MO^*}} \frac{\frac{\partial V_m}{\partial \tau}(X_1, \tau)}{(t_M - \tau)^\alpha} d\tau = 0 \qquad (58)$$

In general, condition (58) is not fullfilled. So, the fractional-order nerve axon model, which uses Caputo fractional order partial derivatives, defined with formula (57), is nonobjective.

In the case of the use of Riemann-Liouville fractional order partial derivatives observer O, instead of equation (54), uses the following equation:

$$c_m^\alpha \cdot \frac{\partial^\alpha V_m}{\partial t_M^\alpha} + \frac{V_m}{r_m} - \frac{1}{r_i} \cdot \frac{\partial^2 V_m}{\partial X_1^2} = I(X_1, t_M)_{appl} \qquad (59)$$

where: $0 < \alpha < 1$ and

$$\frac{\partial^\alpha V_m}{\partial t_M^\alpha}(X_1, t_M) = {}^{[0,t_M]}D^\alpha_{R-L,t_M} V_m(X_1, t_M) = \frac{1}{\Gamma(1-\alpha)} \cdot \frac{\partial}{\partial t_M} \int_0^{t_M} \frac{V_m(X_1, \tau)}{(t_M - \tau)^\alpha} d\tau \qquad (60)$$

The transport description with equations (60) is objective if and only if the following equality holds:

$$\frac{1}{\Gamma(1-\alpha)} \frac{\partial}{\partial t_M} \int_0^{t_{MO^*}} \frac{V_m(X_1,\tau)}{(t_M-\tau)^\alpha} d\tau = 0 \qquad (61)$$

In general, condition (61) is not fullfilled. So, the fractional-order nerve axon model which uses Riemann-Liouville fractional order partial derivatives, defined with formula (60), is nonobjective.

DESCRIPTION OF THE ION TRANSPORT IN THE FRAMEWORK OF FRACTIONAL ORDER NERVE AXON MODEL, WHICH USES GENERAL LIOUVILLE-CAPUTO OR GENERAL RIEMANN-LIOUVILLE FRACTIONAL ORDER PARTIAL DERIVATIVES, DEFINED WITH INTEGRAL REPRESENTATION ON INFINITE INTERVAL, IS OBJECTIVE

If in the framework of the fractional-order nerve axon model, instead of Caputo fractional order partial derivative, general Liouville-Caputo fractional order partial derivative is used, then in case of observer o the transport is described by the equation:

$$c_m^\alpha \cdot \frac{\partial^\alpha V_m}{\partial t_M^\alpha} + \frac{V_m}{r_m} - \frac{1}{r_l} \cdot \frac{\partial^2 V_m}{\partial X_1^2} = I(X_1, t_M)_{appl} \qquad (62)$$

where: $0 < \alpha < 1$ and

$$\frac{\partial^\alpha V_m}{\partial t^\alpha_M}(X_1,t_M) = {}^{(-\infty,t_M]}D^\alpha_{C-L,t_M} V_m(X_1,t_M) =$$
$$= \frac{1}{\Gamma(1-\alpha)} \cdot \int_{-\infty}^{t_M} \frac{\frac{\partial V_m}{\partial \tau}(X_1,\tau)}{(t_M-\tau)^\alpha} d\tau \qquad (63)$$

The transport description with equation (62) is objective, because the following equality holds:

$${}^{(-\infty,t_M]}D^\alpha_{L-C,t_M}{}^\alpha V_m(X_1,t_M) = {}^{(-\infty,t^*_M]}D^\alpha_{L-C,t_M}{}^\alpha V^*_m(X^*_1,t^*_M) \quad (64)$$

If in the framework of the fractional-order nerve axon model, instead of Riemann-Liouville fractional order partial derivative, general Riemann – Liouville fractional order partial derivative is used, then in case of observer o the transport is described by the equation:

$$c^\alpha_m \cdot \frac{\partial^\alpha V_m}{\partial t^\alpha_M} + \frac{V_m}{r_m} - \frac{1}{r_i} \cdot \frac{\partial^2 V_m}{\partial X_1^2} = I(X_1,t_M)_{appl} \qquad (65)$$

where: $0 < \alpha < 1$ and

$$\frac{\partial^\alpha V_m}{\partial t^\alpha_M}(X_1,t_M) = {}^{(-\infty,t_M]}D^\alpha_{R-L,t_M} V_m(X_1,t_M) =$$
$$= \frac{1}{\Gamma(1-\alpha)} \cdot \frac{\partial}{\partial t_M} \int_{-\infty}^{t_M} \frac{V_m(X_1,\tau)}{(t_M-\tau)^\alpha} d\tau \qquad (66)$$

The transport description with equation (65) is objective, because the following equality holds:

$${}^{(-\infty,t_M]}D^\alpha_{R-L,t_M}{}^\alpha V_m(X_1,t_M) = {}^{(-\infty,t^*_M]}D^\alpha_{R-L,t_M}{}^\alpha V^*_m(X^*_1,t^*_M) \quad (67)$$

DESCRIPTION OF THE ION TRANSPORT ACROSS THE ACTIVE AXON MEMBRANE AND ALONG THE AXON, USING INTEGER ORDER PARTIAL DERIVATIVES (ACTIVE NERVE AXON MODEL) IS OBJECTIVE

Combining Hodgkin-Huxley description of ion transport across the axon membrane and ion transport description along the axon a new description, which incorporates both phenomena, is obtained (see [10]. When integer derivatives are used in this new description, the sodium and potassium ions transport across the membrane and along the axon is described, in terms of observer O, with a set of four real valued functions: $V_m = V_m(X_1, t_M)$, $m = m(t_M)$, $h = h(t_M)$, and $n = n(t_M)$. Here: V_m represents the membrane potential; m represents the sodium current activation gating variable; h represents the sodium current inactivation gating variable; n represents the potassium current activation gating variable. These functions verify the following differential equations:

$$C_m \cdot \frac{\partial V_m}{\partial t_M} + \bar{g}_{Na} \cdot m^3 \cdot h \cdot (V_m - V_{Na}) + \bar{g}_K \cdot n^4 \cdot (V_m - V_K) + \bar{g}_L \cdot (V_m - V_L) -$$

$$\frac{1}{r_i} \cdot \frac{\partial^2 V_m}{\partial X_1^2} = I_{appl}(X_1, t_M)$$

$$dn/dt_M = \alpha_n(V_m) \cdot (1-n) - \beta_n(V_m) \cdot n \qquad (68)$$

$$dm/dt_M = \alpha_m(V_m) \cdot (1-m) - \beta_m(V_m) \cdot m$$

$$dh/dt_M = \alpha_h(V_m) \cdot (1-h) - \beta_h(V_m) \cdot h$$

In terms of observer O^* the ion transport across the axon membrane and along the axon axis is described by the set of four functions $V^*_m = V^*_m(X^*_1, t^*_M), m^* = m^*(t^*_M), h^* = h^*(t^*_M), n^* = n^*(t^*_M)$ which verify the differential equations:

$$C_m \cdot \frac{\partial V^*_m}{\partial t^*_m} + \bar{g}_{Na} \cdot m^{*3} \cdot h^* \cdot (V^*_m - V_{Na}) + \bar{g}_K \cdot n^{*4} \cdot (V^*_m - V_K) +$$

$$\bar{g}_L \cdot (V^*_m - V_L) - \frac{1}{r_l} \cdot \frac{\partial^2 V_m}{\partial X_1^{*2}} = I_{appl}^* (X^*_1, t^*_M)$$

$$dn^*/dt^*_M = \alpha_n(V^*_m) \cdot (1 - n^*) - \beta_n(V^*_m) \cdot n^* \tag{69}$$
$$dm^*/dt^*_M = \alpha_m(V^*_m) \cdot (1 - m^*) - \beta_m(V^*_m) \cdot m^*$$
$$dh^*/dt^*_M = \alpha_h(V^*_m) \cdot (1 - h^*) - \beta_h(V^*_m) \cdot h^*$$

Since the applied stimulus current intensity is independent on observer the stimulus current intensity verifies equality

$$I^*(X^*_1, t^*_M)_{appl} = I(X^*_1 + X_{1O^*}, t^*_M + t_{M_{O^*}})_{appl}. \tag{70}$$

Equations (68) and (69) are different, but their solutions describe the same voltage variation under the action of the applied stimulus current. This can be proven showing that if $V_m = V_m(X_1, t_M)$, $m = m(t_M)$, $h = h(t_M)$, $n = n(t_M)$ is a solution of (68), then $V_m^* = V_m^*(X_1^*, t^*_M)$, $m^* = m^*(t^*_M)$, $h^* = h^*(t^*_M)$, $n^* = n^*(t^*_M)$, defined with formulas $V_m^*(X_1^*, t^*_M) = V_m(X^*_1 + X_{1O^*}, t^*_M + t_{M_{O^*}})$, $m^*(t^*_M) = m(t^*_M + t_{M_{O^*}})$, $h^*(t^*_M) = h(t^*_M + t_{M_{O^*}})$ $n^*(t^*_M) = n(t^*_M + t_{M_{O^*}})$ is a solution of (69) and if $V_m^* = V_m^*(X_1^*, t^*_M)$, $m^*(t^*_M), h^*(t^*_M), n^*(t^*_M)$ is a solution of (69), then $V_m = V_m(X_1, t_M)$, $m = m(t_M)$, $h = h(t_M)$, $n = n(t_M)$, defined by $V_m(X_1, t_M) = V_m^*(X_1 + X^*_{1O}, t_M + t^*_{M_O})$, $m(t_M) = m^*(t_M + t^*_{M_O})$, $h(t_M) = h^*(t_M + t^*_{M_O})$, $n(t_M) = n^*(t_M + t^*_{M_O})$, is a solution of (68). So, the description is objective.

Description of the Ion Transport across the Active Axon Membrane and Along the Axon, Using Caputo Fractional Order Partial Derivative or Riemann-Liouville Fractional Order Partial Derivative, Defined with Integral Representation on a Finite Interval (Fractional Order Active Nerve Axon Model) Is Nonobjective

In [10], the author undertakes a study, using for the description of the ion transport across the axon membrane and along the axon partial Caputo or partial Riemann-Liouville fractional order derivatives, defined with integral representation on a finite interval. This type of description in [10] is called fractional order active nerve axon model.

If in the framework of the fractional order active nerve axon model, instead of integer order derivative Caputo fractional order partial derivative, defined with integral representation on finite interval is used, then in case of observer O the transport is described by the equations:

$$C_m \cdot \frac{\partial^\alpha V_m}{\partial t_M^\alpha} + \bar{g}_{Na} \cdot m^3 \cdot h \cdot (V_m - V_{Na}) + \bar{g}_K \cdot n^4 \cdot (V_m - V_K) + \bar{g}_L \cdot (V_m - V_L) -$$

$$\frac{1}{r_i} \cdot \frac{\partial^2 V_m}{\partial X_1^2} = I_{appl}(X_1, t_M)$$

$$dn/dt_M = \alpha_n(V_m) \cdot (1-n) - \beta_n(V_m) \cdot n \qquad (71)$$

$$dm/dt_M = \alpha_m(V_m) \cdot (1-m) - \beta_m(V_m) \cdot m$$

$$dh/dt_M = \alpha_h(V_m) \cdot (1-h) - \beta_h(V_m) \cdot h$$

where: $0 < \alpha < 1$ and

$$\frac{\partial^\alpha V_m}{\partial t_M^\alpha}(X_1,t_M) = {}^{[0,t_M]}D^\alpha_{C,t_M} V_m(X_1,t_M) =$$

$$= \frac{1}{\Gamma(1-\alpha)} \cdot \int_0^{t_M} \frac{\frac{\partial V_m}{\partial \tau}(X_1,\tau)}{(t_M - \tau)^\alpha} d\tau \qquad (72)$$

The transport description with equations (71) is objective if and only if the following equality holds:

$$\frac{1}{\Gamma(1-\alpha)} \int_0^{t_{M^*}} \frac{\frac{\partial V_m}{\partial \tau}(X_1,\tau)}{(t_M - \tau)^\alpha} d\tau = 0 \qquad (73)$$

In general, condition (73) is not fullfilled. So, the fractional order active nerve axon model which uses Caputo fractional order partial derivatives, defined with formula (72), is nonobjective.

If in the framework of the fractional order active nerve axon model, instead of integer order derivative, Riemann-Liouville fractional order partial derivative, defined with integral representation on finite interval is used, then in case of observer O the transport is described by the equations:

$$C_m \cdot \frac{\partial^\alpha V_m}{\partial t_M^\alpha} + \bar{g}_{Na} \cdot m^3 \cdot h \cdot (V_m - V_{Na}) + \bar{g}_K \cdot n^4 \cdot (V_m - V_K) + \bar{g}_L \cdot (V_m - V_L) -$$

$$\frac{1}{r_l} \cdot \frac{\partial^2 V_m}{\partial X_1^2} = I_{appl}(X_1, t_M)$$

$$dn/dt_M = \alpha_n(V_m) \cdot (1-n) - \beta_n(V_m) \cdot n \qquad (74)$$
$$dm/dt_M = \alpha_m(V_m) \cdot (1-m) - \beta_m(V_m) \cdot m$$
$$dh/dt_M = \alpha_h(V_m) \cdot (1-h) - \beta_h(V_m) \cdot h$$

where: $0 < \alpha < 1$ and

$$\frac{\partial^\alpha V_m}{\partial t_M^\alpha}(X_1,t_M) = {}^{[0,t_M]}D^\alpha_{R-L,t_M} V_m(X_1,t_M) = \frac{1}{\Gamma(1-\alpha)} \cdot \frac{\partial}{\partial t_M} \int_0^{t_M} \frac{V_m(X_1,\tau)}{(t_M-\tau)^\alpha} d\tau \qquad (75)$$

The transport description with equations (74) is objective if and only if the following equality holds:

$$\frac{1}{\Gamma(1-\alpha)} \frac{\partial}{\partial t_M} \int_0^{t_{M}*} \frac{V_m(X_1,\tau)}{(t_M-\tau)^\alpha} d\tau = 0 \qquad (76)$$

In general, condition (76) is not fullfilled. So, the fractional order active nerve axon model, which uses Riemann-Liouville fractional order partial derivatives, defined with formula (75), is nonobjective.

DESCRIPTION OF THE ION TRANSPORT IN THE FRAMEWORK OF THE ACTIVE FRACTIONAL ORDER NERVE AXON MODEL, WHICH USES GENERAL LIOUVILLE-CAPUTO OR GENERAL RIEMANN-LIOUVILLE FRACTIONAL ORDER PARTIAL DERIVATIVES, DEFINED WITH INTEGRAL REPRESENTATION ON INFINITE INTERVAL, IS OBJECTIVE

If in the framework of the fractional order active nerve axon model, instead of integer order derivative, general Liouville-Caputo fractional order partial derivative, defined with integral representation on infinite interval is used, then in case of observer O the transport is described by the equations:

$$C_m \cdot \frac{\partial^\alpha V_m}{\partial t^\alpha_M} + \bar{g}_{Na} \cdot m^3 \cdot h \cdot (V_m - V_{Na}) + \bar{g}_K \cdot n^4 \cdot (V_m - V_K) +$$

$$\bar{g}_L \cdot (V_m - V_L) - \frac{1}{r_i} \cdot \frac{\partial^2 V_m}{\partial X_1^2} = I_{appl}(X_1, t_M)$$

$$dn/dt_M = \alpha_n(V_m) \cdot (1-n) - \beta_n(V_m) \cdot n \quad (77)$$

$$dm/dt_M = \alpha_m(V_m) \cdot (1-m) - \beta_m(V_m) \cdot m$$

$$dh/dt_M = \alpha_h(V_m) \cdot (1-h) - \beta_h(V_m) \cdot h$$

where: $0 < \alpha < 1$ and

$$\frac{\partial^\alpha V_m}{\partial t^\alpha_M}(X_1, t_M) = {}^{(-\infty, t_M]}D^\alpha_{L-C, t_M} V_m(X_1, t_M) =$$

$$\frac{1}{\Gamma(1-\alpha)} \cdot \int_{-\infty}^{t_M} \frac{\frac{\partial V_m}{\partial \tau}(X_1, \tau)}{(t_M - \tau)^\alpha} d\tau \quad (78)$$

The transport description with equations (77) is objective, because the following equality holds:

$${}^{(-\infty, t_M]}D^\alpha_{L-C, t_M} V_m(X_1, t_M) = {}^{(-\infty, t_M]}D^\alpha_{L-C, t_M} V^*_m(X^*_1, t^*_M) \quad (79)$$

If in the framework of the fractional order active nerve axon model, instead of integer order derivative, Riemann-Liouville fractional order partial derivative, defined with integral representation on infinite interval is used, then in case of observer O the transport is described by the equations:

$$C_m \cdot \frac{\partial^\alpha V_m}{\partial t^\alpha_M} + \bar{g}_{Na} \cdot m^3 \cdot h \cdot (V_m - V_{Na}) + \bar{g}_K \cdot n^4 \cdot (V_m - V_K) +$$

$$\bar{g}_L \cdot (V_m - V_L) - \frac{1}{r_i} \cdot \frac{\partial^2 V_m}{\partial X_1^2} = I_{appl}(X_1, t_M)$$

$$dn/dt_M = \alpha_n(V_m) \cdot (1-n) - \beta_n(V_m) \cdot n \qquad (80)$$
$$dm/dt_M = \alpha_m(V_m) \cdot (1-m) - \beta_m(V_m) \cdot m$$
$$dh/dt_M = \alpha_h(V_m) \cdot (1-h) - \beta_h(V_m) \cdot h$$

where: $0 < \alpha < 1$ and

$$\frac{\partial^\alpha V_m}{\partial t_M^\alpha}(X_1, t_M) = {}^{(-\infty, t_M]}D^\alpha_{R-L, t_M} V_m(X_1, t_M) = \frac{1}{\Gamma(1-\alpha)} \cdot \frac{\partial}{\partial t_M} \int_{-\infty}^{t_M} \frac{V_m(X_1, \tau)}{(t_M - \tau)^\alpha} d\tau \qquad (81)$$

The transport description with equations (80) is objective, because the following equality holds:

$${}^{(-\infty, t_M]}D^\alpha_{R-L, t_M} V_m(X_1, t_M) = {}^{(-\infty, t_M]}D^\alpha_{R-L, t_M} V^*_m(X^*_1, t^*_M) \qquad (82)$$

DESCRIPTION OF THE ION TRANSPORT IN A NEURAL NETWORK USING FRACTIONAL ORDER DERIVATIVES, HAVING INTEGRAL REPRESENTATION ON FINITE INTERVAL, IS NONOBJECTIVE

In [10] a network of $N = 50$ randomly connected fractional order Hodgkin-Huxley type neurons is presented.

In the case of Caputo fractional order derivative, having integral representation on a finite interval, to the fractional order Hodgkin-Huxley description a synaptic current is added such that the dynamics of the ith neuron in the observer O description is governed by the following equations:

$$C_m^\alpha {}^{[0,t_M]} D_C^\alpha V_m^i(t_M) + \bar{g}_{Na} \cdot m_i^3 \cdot h_i \cdot (V_m^i - V_{Na}) +$$
$$\bar{g}_K \cdot n_i^4 \cdot (V_m^i - V_K) + \bar{g}_L \cdot (V_m^i - V_L) = I_{appl}^i(t_M) - I_{syn,i} \quad (83)$$
$$dn_i/dt_M = \alpha_n(V_m^i) \cdot (1 - n_i) - \beta_n(V_m^i) \cdot n_i$$
$$dm_i/dt_M = \alpha_m(V_m^i) \cdot (1 - m_i) - \beta_m(V_m^i) \cdot m_i$$
$$dh_i/dt_M = \alpha_h(V_m^i) \cdot (1 - h_i) - \beta_h(V_m^i) \cdot h_i$$

The synaptic current is the sum of the excitatory and inhibitory synaptic currents:

$$I_{syn,i} = I_{synE,i} + I_{synI,i} \quad (84)$$

where:

$$I_{synE,i} = g_{syn}(\sum_{j \in S_{ex}} s_{ji}(V_m^j))(V_m^i - E_{syn,ex}) \quad (85)$$

$$I_{synI,i} = g_{syn}(\sum_{j \in S_{in}} s_{ji}(V_m^j))(V_m^i - E_{syn,in}) \quad (86)$$

S_{ex} and S_{in} are the presynaptic neurons with connection to neuron i with excitatory and inhibitory, respectively, synapses; and s_{ji} is the gating variable for the post synaptic conductance, an instantaneous, sigmoidal function of presynaptic cell potential V_m^j with a threshold V_{syn}. That is:

$$s_{ji}(V_m^j) = \frac{1}{1 + \exp[-(V_m^j - V_{syn})/k_{syn}]} \quad (87)$$

The synaptic current parameters are given in [10].
The transport description with equations (83) is objective, if and only if the following equality holds:

$$\frac{1}{\Gamma(1-\alpha)} \int_0^{t_{MO^*}} \frac{\frac{dV_m^i}{d\tau}(\tau)}{(t_M - \tau)^\alpha} d\tau = 0 \tag{88}$$

In general, condition (88) is not fullfilled. So, the ion transport in neural network using Caputo fractional order derivative, defined with formula (11), is nonobjective.

In the case of Riemann-Liouville fractional order derivative, having integral representation on a finite interval, the dynamics of the ith neuron in the observer O description is governed by equations similar to the equations (83)-(87) except the first equations from the system (83), where instead of the Caputo fractional order derivative ((11)), Riemann-Liouville fractional order derivative ((12)) appears. In this case, the transport description is objective if and only if the following equality holds:

$$\frac{1}{\Gamma(1-\alpha)} \cdot \frac{d}{dt_M} \int_0^{t_{MO^*}} \frac{V_m^i(\tau)}{(t_M - \tau)^\alpha} d\tau = 0 \tag{89}$$

In general, condition (89) is not fullfilled. So, the ion transport in neural network using Riemann-Liouville fractional order derivative, defined with formula (12), is nonobjective.

DESCRIPTION OF THE ION TRANSPORT IN A NEURAL NETWORK USING FRACTIONAL ORDER DERIVATIVES, HAVING INTEGRAL REPRESENTATION ON INFINITE INTERVAL, IS OBJECTIVE

In the case of Liouville-Caputo fractional order derivative, having integral representation on infinite interval, the dynamics of the ith neuron, in the observer O description, is governed by equations similar to the equations (83)-(87), except the first equations from the system (83) where instead of the Caputo fractional order derivative ((11)), Liouville-Caputo fractional order derivative

((13)), appears. In this case the transport description is objective because the following equality holds:

$$^{(-\infty,t_M]}D^\alpha_{L-C}V^i_m(t_M) = {}^{(-\infty,t_M]}D^\alpha_{L-C}V*_m(t*_M) \qquad (90)$$

In the case of Riemann-Liouville fractional order derivative, having integral representation on infinite interval, the dynamics of the ith neuron, in the observer O description, is governed by equations similar to the equations (83)-(87), except the first equations from the system (83) where instead of the Caputo fractional order derivative ((11)), Riemann-Liouville fractional order derivative ((14)), appears. In this case the transport description is objective, because the following equality holds:

$$^{(-\infty,t_M]}D^\alpha_{R-L}V_m(t_M) = {}^{(-\infty,t_M]}D^\alpha_{R-L}V*_m(t*_M) \qquad (91)$$

DESCRIPTION OF THE ION TRANSPORT IN A NEURAL NETWORK USING INTEGER ORDER DERIVATIVE IS OBJECTIVE

In case of integer order derivative the dynamics of the ith neuron, in the observer O description is governed by equations similar to the equations (83)-(87), except the first equations from the system (83) where instead of the Caputo fractional order derivative, first order derivative appears. In this case the transport description is objective, because the following equality holds:

$$\frac{dV^i_m}{dt_M} = \frac{dV*^i_m}{dt*_M} \qquad (92)$$

Conclusion

1. Electron transport description in electric circuit with fractional order Caputo derivative (11) is nonobjective, but electron transport description in electric circuit with fractional order general Liouville-Caputo derivative (13), or integer order derivative, is objective.
2. Electron transport description in electric circuit with fractional order Riemann-Liouville derivative (12) is nonobjective, but electron transport description in electric circuit with fractional order general Riemann-Liouville derivative (14) or integer order derivative is objective.
3. Hodgkin-Huxley type description of potasium and sodium ion transport across the neuron axon membrane with fractional order Caputo derivative (11) is nonobjective, but the same type of description with fractional order general Liouville-Caputo derivative (13), or integer order derivative, is objective.
4. Hodgkin-Huxley type description of potasium and sodium ion transport across the neuron axon membrane with fractional order Riemann-Liouville derivative (12) is nonobjective, but the same type of description with fractional order general Riemann-Liouville derivative (14), or integer order derivative, is objective.
5. Hodgkin-Huxley type description of potasium and sodium ion transport across the neuron axon membrane and along the axon with fractional order Caputo derivative (11) is nonobjective, but the same type of description with fractional order general Liouville-Caputo derivative (13), or integer order derivative, is objective.
6. Hodgkin-Huxley type description of potasium and sodium ion transport across the neuron axon membrane and along the axon with fractional order Riemann-Liouville derivative (12) is nonobjective, but the same type of descriptions with fractional order general Riemann-Liouville derivative (14), or integer order derivative, is objective.
7. Morris-Lecar type description of calcium and potassium ion transport across the neuron axon membrane with fractional order Caputo derivatives (11) is nonobjective, but the same type of descriptions with

fractional order general Liouville-Caputo derivative (13), or integer order derivative, is objective.

8. Morris-Lecar type description of calcium and potasium ion transport across the neuron axon membrane with fractional order Riemann-Liouville derivative (12) is nonobjective, but the same type of descriptions with fractional order general Riemann-Liouville derivative (14), or integer order derivative, is objective.

9. Ion transport in randomly connected Hodgkin-Huxley type neural networks, described with fractional order Caputo derivatives (11), is nonobjective, but the same type of description with fractional order general Liouville-Caputo derivative (13), or integer order derivative, is objective.

10. Ion transport in randomly connected Hodgkin-Huxley type neural networks, described with fractional order Riemann-Liouville derivative (12), is nonobjective, but the same type of description with fractional order general Riemann-Liouville derivative (14), or integer order derivative, is objective.

REFERENCES

[1] Debs, Talal A. & Redhead, Michael L. G. (2007). *Objectivity, Invariance, and Convention: Symmetry in Physical Science*. Cambridge: Harvard University Press.

[2] Balabanian, Norman. & Bickart, Theodore A. (1969). *Electrical network theory*, (John Wiley & Sons, Inc.,), pp. 30-85.

[3] Hirsch, Morris W. & Smale, Stephen S. (1974). *Differential equations, dynamical systems and linear algebra* (Academic Press, New York and London), pp. 211-238.

[4] Curie, Jacques. (1889). "Research on the conductivity of crystalline materials" *Ann. de Chimie et de la Physique*, *18*, 203-269.

[5] v Schweidler, Egon Ritter. (1907). "Studies concerning some anomalies in the behavior of dielectrics" Aus den Sitzungsber. der kaiserl. *Akad. der Wissensch. in Wien. Matem. Naturw.*, *116*, 1055-1060.

[6] Caputo, Michele. (1967). "Linear model of dissipation whose Q is almost frequency independent-II." *Geophysical Journal International.*, *13* (5), 529–539.

[7] Demirci, Elif. & Ozalp, Nuri. (2012). "A method for solving differential equations of fractional order." *Journal of Computational and Applied Mathematics*, *236*, 2754-2762.

[8] Ortigueira, Manuel D. & Machado, Jose A. Tenreiro. (2017). "Which Derivative?" *Fractal and Fractional.*, *1*, 3, doi:10.3390/fractalfract1010003.

[9] Moreles, Miguel A. & Lainez, Rafael. (2016). *Mathematical Modelling of Fractional Order Circuits.* arXiv:1602.03541 [physics class-ph.].

[10] Weinberg, Seth H. (2015). "Membrane Capacitive Memory Alters Spiking in Neurons Described by the Fractional Order Hodgkin-Huxley Model" *PloS ONE*, *10*(5), e0126629.

[11] Hodgkin, Alan Lloyd. & Huxley, Andrew Fielding. (1952). "A Quantitative Description of Membrane Current and its Application to Conduction and Excitation in Nerve." *J. Physiol.* (Lond.), *117*, 500-544.

[12] Morris, Catherine. & Lecar, Harold. (1981). "Voltage oscillations in the barnacle giant muscle fiber". *Biophysical J.*, *35*, 193-213.

[13] Keynes, Richard., Rojas, D., Eduardo, Taylor., Robert, E. & Vergara, Jorge, R. (1973). "Calcium and potassium systems of a giant barnacle muscle fiber under membrane potential control". *J. Physiol.* (Lond.), *229*, 409-455.

[14] Hagiwara, Susumu., Hayashi, Hideo. & Takahashi, Kunitaro. (1969). "Calcium and potassium currents of the membrane of a barnacle muscle fibre in relation to the calcium spike." *J. Physiol.*, *205*, 115-129.

[15] Hagiwara, Susumu., Fukuda, Jun. & Eaton, Douglas, C. (1974). "Membrane currents carried by Ca, Sr, and Ba in barnacle muscle fibre during voltage clamp." *J. Gen. Physiol.*, *63*, 564-578.

[16] Murayama, K. & Lakshminarayanaiah, N. (1977). "Some electrical properties of the membrane of the barnacle muscle fibres under internal perfusion". *J. Membr. Biol.*, *35*, 257-283.

[17] Beirao, Paulo S. & Lakshminarayanaiah, N. (1979). "Calcium carrying system in the giant muscle fibre of the barnacle species Balanus Nubilus". *J. Physiol.*, *293*, 319-327.

[18] Hagiwara, Susumu. & Nakajima, Shigehiro. (1966). "Effects of the intracellular Ca ion concentration upon excitability of the muscle fiber membrane of a barnacle" *J. Gen. Physiol.*, *49*, 807-818.

[19] Brandibur, Oana. & Kaslik, Eva. (2017). "Stability properties of a two dimensional system involving one Caputo derivative and applications to the investigation of a fractional order Morris-Lecar neuronal model" *Nonlinear Dyn.*, *90*, 2371-2386.

In: Understanding Time Evolution
Editor: Asger S. Thorsen

ISBN: 978-1-53617-874-6
© 2020 Nova Science Publishers, Inc.

Chapter 4

OBJECTIVE AND NONOBJECTIVE MATHEMATICAL DESCRIPTION OF THE MECHANICAL MOVEMENT OF A MATERIAL POINT, DUE TO THE USE OF DIFFERENT TYPE OF FRACTIONAL ORDER DERIVATIVES

Agneta M. Balint[1] *and Stefan Balint*[2,*]

[1]Department of Physics, West University of Timisoara,
Timisoara, Romania
[2]Department of Computer Science, West University of Timisoara,
Timisoara, Romania

ABSTRACT

In this chapter it is shown that, in the material point mechanics the mathematical description using Caputo or Riemann-Liouville fractional order derivative defined with integral representation on finite interval is nonobjective and, the mathematical description using general Caputo Liouville or general Riemann-Liouville fractional order derivative defined with integral representation on infinite interval, is objective.

---

[*] Corresponding Author's Email: stefan.balint@e-uvt.ro (Corresponding author).

Keywords: objectivity of a mathematical description, mechanical movement description, material point, fractional order derivative

INTRODUCTION

The concept of objectivity in science means that qualitative and quantitative descriptions of a certain phenomenon remain unchanged when the phenomena is observed by different observers; that is, it is possible to reconcile observations of the process into a single coherent description of it [1].

Galileo Galilee (1564-1642) said "The mechanical event is independent from the observer. For frames moving uniformly with respect to each other, both states are mechanically equivalent."

Isaac Newton (1643- 1727) said "The mechanical event is independent from the observer. This holds also for accelerated systems if the frames of references are fixed with respect to absolute space (with respect to the fixed stars)."

Albert Einstein (1879-1955) said "The mechanical event is independent from the observer. There is no special reference point. The same holds for accelerated systems (general relativity). Even more, if the theory is subjected to relativity, it should be generally covariant under all transformations, not just rigid body motions."

SOME EXAMPLES OF OBJECTIVE AND NONOBJECTIVE MATHEMATICAL DESCRIPTIONS

Objective Description of the Movement of a Material Particle

In the following, what means objectivity of the classical description of the movement of a material point is illustrated [2]. In classical mechanics, observer O represents a material particle by a point P, called a material point in the three-dimensional affine Euclidian space E_3. To describe the position of the material point, observer O chooses a fixed orthogonal reference frame

$R_O = (O; \vec{e}_1, \vec{e}_2, \vec{e}_3)$ in E_3 and describes the position of P using its coordinates with respect to the reference frame R_O. To describe the time evolution, observer O chooses a moment of time M_O for fixing the origin of the time measurement and a unit [second] for the time measurement. A moment of time M which is earlier than M_O is represented by a negative real number $t_M < 0$, a moment of time M which is later than M_O is represented by a positive real number $t_M > 0$ and the moment of time M_O is represented by the real number $t_{M_O} = 0$. At any moment of time M, represented by t_M, the observer considers the coordinate $(X_1(t_M), X_2(t_M), X_3(t_M))$ of the material point at the moment of time t_M with respect to the reference frame R_O and describes the movement of the material point with the set of the real functions $X_1(t_M), X_2(t_M), X_3(t_M)$. A second observer O^*, uses a similar procedure and describes the same movement of the material point with the set of real functions $X^*_1(t^*_M), X^*_2(t^*_M), X^*_3(t^*_M)$, representing the coordinates of the material point with respect to a second fixed reference frame $R_{O^*} = (O^*; \vec{e}^*_1, \vec{e}^*_2, \vec{e}^*_3)$. For the observer O^* the origin of the time measurement is M_{O^*}, the unit is [second]; a moment of time M, which is earlier than M_{O^*}, is represented by a negative number $t^*_M < 0$; the moment of time M_{O^*} is represented by the real number $t^*_{M_{O^*}} = 0$ and a moment of time M, which is later than M_{O^*}, is represented by a positive real number $t^*_M > 0$. In case of the observer O, a moment of time M is described by the real number t_M and in case of the observer O^* by the real number t^*_M. For the numbers t_M and t^*_M the following relations hold:

$$t_M = t^*_M + t_{M_{O^*}} \qquad (1)$$

$$t*_M = t_M + t*_{M_O} \tag{2}$$

In the above relations $t_{M_{O^*}}$ is the real number which represents the moment M_{O^*} in the system of time measuring of the observer O and $t*_{M_O}$ is the real number which represents the moment of time M_O in time measuring system of observer O^*.

For any moment of time M, the coordinates $(X_1(t_M), X_2(t_M), X_3(t_M))$ with respect to R_O and $(X*_1(t*_M), X*_2(t*_M), X*_3(t*_M))$ with respect to R_{O^*} represent points in the three dimensional affine Euclidian space E_3. These points have to coincide with the material point position at the moment of time M. Therefore, for the coordinates the following relations hold:

$$X_k(t_M) = X_{kO^*} + \sum_{i=1}^{i=3} a_{ik} X*_i(t*_M) \quad k = 1,2,3$$
$$t_M = t*_M + t_{M_{O^*}} \tag{3}$$

or equivalently

$$X*_k(t_M) = X*_{kO} + \sum_{i=1}^{i=3} a_{ki} X_i(t_M) \quad k = 1,2,3$$
$$t*_M = t_M + t*_{M_O} \tag{4}$$

The significance of the quantities appearing in the above relations are: $a_{ij} = \langle \vec{e}*_i, \vec{e}_j \rangle = $ constant = scalar product of the unit vectors $\vec{e}*_i$ and $\vec{e}_j$ in E_3 i.e.,

$$\vec{e}*_i = \sum_{k=1}^{k=3} a_{ik} \vec{e}_k \quad \vec{e}_i = \sum_{k=1}^{k=3} a_{ki} \vec{e}*_k \tag{5}$$

$(X_{1O*}, X_{2O*}, X_{3O*})$ are the coordinates of the point $O*$ with respect to the reference frame R_O, $(X*_{1O}, X*_{2O}, X*_{3O})$ are the coordinates of the point O with respect to the reference frame R_{O*}.

Relations (3) and (4) reconcile the description made by the two observers and make possible the description of the movement by the set of functions $X_1(t_M), X_2(t_M), X_3(t_M)$ or by the set of functions $X*_1(t*_M), X*_2(t*_M), X*_3(t*_M)$. This means that the above presented movement description is objective.

Caputo, Riemann-Liouville Fractional Order Derivatives Defined with Integral Representation on Finite Interval

According to [3], for a continuously differentiable function $f:[0,\infty) \to R$ the Caputo fractional derivative (C) of order α, $(0 < \alpha < 1)$ is defined with formula:

$$^{[0,t]}D_C^\alpha f(t) = \frac{1}{\Gamma(1-\alpha)} \cdot \int_0^t \frac{f'(\tau)}{(t-\tau)^\alpha} d\tau \qquad (6)$$

Remark that the derivative defined with (6) was considered by other people before, like Gherasimov (see [4]). So, the name of "Caputo," given in this chapter, maybe is not appropriate.

For a continuously differentiable function $f:[0,\infty) \to R$ the Riemann-Liouville (R-L) fractional derivative of order α, $(0 < \alpha < 1)$, according to [3], is defined with formula:

$$^{[0,t]}D_{R-L}^\alpha f(t) = \frac{1}{\Gamma(1-\alpha)} \cdot \frac{d}{dt} \int_0^t \frac{f(\tau)}{(t-\tau)^\alpha} d\tau \qquad (7)$$

According to [4], the derivative having integral representation on infinite interval was proposed by Liouville in 1832. Some people call it Liouville-Caputo derivative.

General Liouville-Caputo, General Riemann-Liouville Fractional Order Derivatives Defined with Integral Representation on Infinite Interval

For a continuously differentiable function $f:(-\infty,\infty) \to R$ the general Liouville-Caputo (L-C) and the general Riemann-Liouville fractional order derivative of order α, $(0 < \alpha < 1)$ according to [4], are defined with formula:

$$^{(-\infty,t]}D_{L-C}^{\alpha} f(t) = \frac{1}{\Gamma(1-\alpha)} \cdot \int_{-\infty}^{t} \frac{f'(\tau)}{(t-\tau)^{\alpha}} d\tau \qquad (8)$$

$$^{(-\infty,t]}D_{R-L}^{\alpha} f(t) = \frac{1}{\Gamma(1-\alpha)} \cdot \frac{d}{dt} \int_{-\infty}^{t} \frac{f(\tau)}{(t-\tau)^{\alpha}} d\tau \qquad (9)$$

where Γ is the Euler gamma function.

Objective Description of the Material Point Velocity Using First Order Derivative

For observer O the velocity $\vec{V}_M$ at the moment of time M is the vector in E_3 obtained translating the vector $\vec{X}'(t_M) = \sum_{i=1}^{i=3} X'_i(t_M) \vec{e}_i$ in the point of coordinates $(X_1(t_M), X_2(t_M), X_3(t_M))$. Here $X'_i(t_M)$ is the first order derivative of the function $X_i(t_M)$ at t_M for $i = 1,2,3$.

Objective and Nonobjective Mathematical Description of ...

For observer O^* the velocity $\vec{V}^*_M$ at the moment of time M is the vector in E_3 obtained translating the vector $\vec{X}^{*\prime}(t^*_M) = \sum_{i=1}^{i=3} X^{*\prime}_i(t^*_M)\vec{e}^*_i$ in the point of coordinates $(X^*_1(t^*_M), X^*_2(t^*_M), X^*_3(t^*_M))$. Here $X^{*\prime}_i(t^*_M)$ is the first order derivative of the function $X^*_i(t^*_M)$ at t^*_M for $i = 1,2,3$.

The velocity description is objective if and only if $\vec{V}_M = \vec{V}^*_M$. Equality $\vec{V}_M = \vec{V}^*_M$ can be proven by differentiating in (3) and obtaining:

$$X'_k(t_M) = \sum_{i=1}^{i=3} a_{ik} X^{*\prime}_i(t^*_M) \qquad k = 1,2,3 \tag{10}$$

Equalities (10) show that equality $\vec{V}_M = \vec{V}^*_M$ holds. So, we obtain that velocity described by first order derivatives is objective.

Objective Description of the Material Point Velocity Using General Caputo-Liouville Fractional Order Derivative Defined with Formula (8)

For observer O the general Liouville-Caputo velocity of order α, $^{L-C}\vec{V}^\alpha_M$, at the moment of time M is the vector in E_3 obtained translating the vector

$$^{(-\infty,t_M]}D^\alpha_{L-C}\vec{X}(t_M) = \sum_{i=1}^{i=3} {}^{(-\infty,t_M]}D^\alpha_{L-C}X_i(t_M)\vec{e}_i \tag{11}$$

in the point of coordinates $(X_1(t_M), X_2(t_M), X_3(t_M))$. Here $^{(-\infty,t_M]}D^\alpha_{L-C}X_i(t_M)$ is the general Liouville-Caputo α order derivative of the function $X_i(t_M)$ at t_M for $i = 1,2,3$ (formula (8)).

For observer O^* the general Liouville-Caputo velocity of order α, $^{L-C}\vec{V}*^\alpha{}_M$, at the moment of time M is the vector in E_3 obtained translating the vector

$$^{(-\infty,t^*_M]}D^\alpha{}_{L-C}\vec{X}*(t^*_M) = \sum_{i=1}^{i=3} {}^{(-\infty,t^*_M]}D^\alpha{}_{L-C}X^*_i(t^*_M)\vec{e}^*_i \quad (12)$$

in the point of coordinates $(X^*_1(t^*_M), X^*_2(t^*_M), X^*_3(t^*_M))$.

Here $^{(-\infty,t^*_M]}D^\alpha{}_{L-C}X^*_i(t^*_M)$ is the general Liouville-Caputo α order derivative of the function $X^*_i(t^*_M)$ at t^*_M for $i=1,2,3$ (formula (8)).

The general Liouville-Caputo velocity of order α is objective if and only if equality $^{L-C}\vec{V}^\alpha{}_M = {}^{L-C}\vec{V}*^\alpha{}_M$ holds. Equality $^{L-C}\vec{V}^\alpha{}_M = {}^{L-C}\vec{V}*^\alpha{}_M$ can be proven showing first equalities:

$$^{(-\infty,t_M]}D^\alpha{}_{L-C}X_i(t_M) = {}^{(-\infty,t^*_M]}D^\alpha{}_{L-C}X^*_i(t^*_M) \text{ for } i=1,2,3. \quad (13)$$

For (13) remark the following equalities:

$$\begin{aligned}
{}^{(-\infty,t_M]}D_{L-C}^\alpha X_i(t_M) &= \frac{1}{\Gamma(1-\alpha)} \cdot \int_{-\infty}^{t_M} \frac{X_k'(\tau)}{(t_M-\tau)^\alpha} d\tau = \\
&= \frac{1}{\Gamma(1-\alpha)} \cdot \int_{-\infty}^{t_M-t_{M_{O^*}}} \frac{X_i'(t_{M_{O^*}}+\xi)}{(t_M-t_{M_{O^*}}-\xi)^\alpha} d\xi = \\
&= \frac{1}{\Gamma(1-\alpha)} \cdot \int_{-\infty}^{t^*_M} \frac{X_i'(t_{M_{O^*}}+\xi)}{(t^*_M-\xi)^\alpha} d\xi = \\
&= \frac{1}{\Gamma(1-\alpha)} \cdot \int_{-\infty}^{t^*_M} \frac{X^*_i{'}(\xi)}{(t^*_M-\xi)^\alpha} d\xi = {}^{(-\infty,t^*_M]}D_{L-C}^\alpha X^*_i(t^*_M).
\end{aligned} \quad (14)$$

Now from (11), (12), (13) we deduce the following equalities:

$$^{(-\infty,t_M]}D^\alpha{}_{L-C}\vec{X}(t_M) = \sum_{i=1}^{i=3} {}^{(-\infty,t_M]}D^\alpha{}_{L-C}X_i(t_M)\vec{e}_i =$$

$$\sum_{i=1}^{i=3} {}^{(-\infty,t^*_M]}D^\alpha{}_{L-C}X^*_i(t^*_M)\vec{e}_i =$$

$$\sum_{i=1}^{i=3} {}^{(-\infty,t^*_M]}D^\alpha{}_{L-C}X^*_i(t^*_M)\sum_{k=1}^{k=3} a_{ki}\vec{e}^*_k = \qquad (15)$$

$$\sum_{k=1}^{k=3}(\sum_{i=1}^{i=3} a_{ki} \cdot {}^{(-\infty,t^*_M]}D^\alpha{}_{L-C}X^*_i(t^*_M))\vec{e}^*_k =$$

$$^{(-\infty,t^*_M]}D^\alpha{}_{L-C}\vec{X}^*(t^*_M)$$

So, we obtain that the general Liouville-Caputo velocity of order α is objective.

Objective Description of the Material Point Velocity Using General Riemann-Liouville Fractional Order Derivative Defined with Formula (9)

For observer O the general Riemann-Liouville velocity ${}^{R-L}\vec{V}^\alpha{}_M$ of order α, at the moment of time M, is the vector obtained translating in E_3 the vector

$$^{(-\infty,t_M]}D^\alpha{}_{R-L}\vec{X}(t_M) = \sum_{i=1}^{i=3} {}^{(-\infty,t_M]}D^\alpha{}_{R-L}X_i(t_M)\vec{e}_i \qquad (16)$$

to the point of coordinates $(X_1(t_M), X_2(t_M), X_3(t_M))$. Here ${}^{(-\infty,t_M]}D^\alpha{}_{R-L}X_i(t_M)$ is the general Riemann-Liouville α order derivative of the function $X_i(t_M)$ at t_M for $i = 1,2,3$ ((9)).

For observer O^* the general Riemann-Liouville velocity ${}^{R-L}\vec{V}^{*\alpha}{}_M$ of order α at the moment of time M is the vector obtained translating in E_3 the vector

$$^{(-\infty,t^*_M]}D^\alpha_{R-L}\vec{X}*(t^*_M) = \sum_{i=1}^{i=3} {}^{(-\infty,t^*_M]}D^\alpha_{R-L}X^*_i(t^*_M)\vec{e}^*_i \quad (17)$$

to the point of coordinates $(X^*_1(t^*_M), X^*_2(t^*_M), X^*_3(t^*_M))$.

Here $^{(-\infty,t^*_M]}D^\alpha_{R-L}X^*_i(t^*_M)$ is the general Riemann-Liouville α order derivative of the function $X^*_i(t^*_M)$ at t^*_M for $i=1,2,3$ (formula (9)).

The general Riemann-Liouville velocity of order α is objective if and only if equality ${}^{R-L}\vec{V}^\alpha_M = {}^{R-L}\vec{V}*^\alpha_M$ holds. This equality can be proven showing first equalities:

$$^{(-\infty,t_M]}D^\alpha_{R-L}X_i(t_M) = {}^{(-\infty,t^*_M]}D^\alpha_{R-L}X^*_i(t^*_M) \quad \text{for } i=1,2,3. \quad (18)$$

For (18) remark equalities:

$$\begin{aligned}
{}^{(-\infty,t_M]}D^\alpha_{R-L}X_i(t_M) &= \frac{1}{\Gamma(1-\alpha)} \cdot \frac{d}{dt_M} \int_{-\infty}^{t_M} \frac{X_i(\tau)}{(t_M-\tau)^\alpha} d\tau = \\
&= \frac{1}{\Gamma(1-\alpha)} \cdot \frac{d}{dt_M} \int_{-\infty}^{t_M - t_{M_{O^*}}} \frac{X_i(t_{M_{O^*}}+\xi)}{(t_M - t_{M_{O^*}} - \xi)^\alpha} d\xi = \\
&= \frac{1}{\Gamma(1-\alpha)} \cdot \frac{d}{dt_M *} \int_{-\infty}^{t^*_M} \frac{X_i(t_{M_{O^*}}+\xi)}{(t^*_M - \xi)^\alpha} d\xi = \quad (19)\\
&= \frac{1}{\Gamma(1-\alpha)} \cdot \frac{d}{dt_M *} \int_{-\infty}^{t^*_M} \frac{X^*_i(\xi)}{(t^*_M - \xi)^\alpha} d\xi = \\
&= {}^{(-\infty,t^*_M]}D^\alpha_{R-L}X^*_i(t^*_M)
\end{aligned}$$

Now, from (16), (17), (18) it follows that the following equalities holds:

$$\begin{aligned}
^{(-\infty,t_M]}D^\alpha_{R-L}\vec{X}(t_M) &= \sum_{i=1}^{i=3} {}^{(-\infty,t_M]}D^\alpha_{R-L}X_i(t_M)\vec{e}_i = \\
&\sum_{i=1}^{i=3} {}^{(-\infty,t^*_M]}D^\alpha_{R-L}X^*_i(t^*_M)\vec{e}_i = \\
&\sum_{i=1}^{i=3} {}^{(-\infty,t^*_M]}D^\alpha_{R-L}X^*_i(t^*_M)\cdot\sum_{k=1}^{k=3}a_{ki}\vec{e}^*_k = \qquad (20) \\
&\sum_{k=1}^{k=3}\left(\sum_{i=1}^{i=3}a_{ki}{}^{(-\infty,t^*_M]}D^\alpha_{R-L}X^*_i(t^*_M)\right)\vec{e}^*_k = \\
&{}^{(-\infty,t^*_M]}D^\alpha_{R-L}\vec{X}^*(t^*_M)
\end{aligned}$$

So, we obtain that the general Riemann-Liouville velocity of order α is objective.

Nonobjective Description of the Material Point Velocity Using Caputo Fractional Order Derivative Defined with Formula (6)

For observer O the Caputo velocity ${}^C\vec{V}^\alpha_M$ of order α, at the moment of time M is the vector obtained translating in E_3 the vector

$${}^{[0,t_M]}D^\alpha_C\vec{X}(t_M) = \sum_{i=1}^{i=3} {}^{[0,t_M]}D^\alpha_C X_i(t_M)\vec{e}_i \quad (21)$$

to the point of coordinates $(X_1(t_M), X_2(t_M), X_3(t_M))$. Here ${}^{[0,t_M]}D^\alpha_C X_i(t_M)$ is the Caputo derivative of order α of the function $X_i(t_M)$ at t_M for $i=1,2,3$ (formula (6)).

For observer O^* the Caputo velocity ${}^C\vec{V}^{*\alpha}_M$ of order α at the moment of time M is the vector obtained translating in E_3 the vector

$${}^{[0,t^*_M]}D^\alpha_C\vec{X}^*(t^*_M) = \sum_{i=1}^{i=3} {}^{[0,t^*_M]}D^\alpha_C X^*_i(t^*_M)\vec{e}^*_i \quad (22)$$

to the point of coordinates $(X^*_1(t^*_M), X^*_2(t^*_M), X^*_3(t^*_M))$. Here ${}^{[0,t^*_M]}D^\alpha_C X^*_i(t^*_M)$

is the Caputo derivative of order α of the function $X^*_i(t^*_M)$ at t^*_M for $i = 1,2,3$ (formula (6)).

The above description of Caputo velocity of order α is objective if and only if equality ${}^C\vec{V}^\alpha{}_M = {}^C\vec{V}*^\alpha{}_M$ holds.

Equality ${}^C\vec{V}^\alpha{}_M = {}^C\vec{V}*^\alpha{}_M$ holds if and only if:

$$^{[0,t_M]}D^\alpha{}_C\vec{X}(t_M) = \sum_{i=1}^{i=3} {}^{[0,t_M]}D^\alpha{}_C X_i(t_M)\vec{e}_i = {}^{[0,t^*_M]}D^\alpha{}_C\vec{X}*(t^*_M) = \sum_{i=1}^{i=3} {}^{[0,t^*_M]}D^\alpha{}_C X^*_i(t^*_M)\vec{e}*_i \qquad (23)$$

Hence, by using $\vec{e}*_i = \sum_{k=1}^{k=3} a_{ik}\vec{e}_k$, the following equalities are obtained:

$$\sum_{i=1}^{i=3} {}^{[0,t_M]}D^\alpha{}_C X_i(t_M)\vec{e}_i = \sum_{i=1}^{i=3} {}^{[0,t^*_M]}D^\alpha{}_C X^*_i(t^*_M)\vec{e}*_i =$$

$$= \sum_{i=1}^{i=3} {}^{[0,t^*_M]}D^\alpha{}_C X_i*(t^*_M) \cdot \left(\sum_{k=1}^{k=3} a_{ik}\vec{e}_k\right) = \qquad (24)$$

$$\sum_{k=1}^{k=3}\left(\sum_{i=1}^{i=3} a_{ik} \cdot {}^{[0,t^*_M]}D^\alpha{}_C X^*_i(t^*_M)\right)\vec{e}_k$$

So, Caputo velocity of order α, defined by formula (21), is objective if and only if the next equalities hold:

$$^{[0,t_M]}D^\alpha{}_C X_k(t_M) = \sum_{i=1}^{i=3} a_{ik} \cdot {}^{[0,t^*_M]}D^\alpha{}_C X^*_i(t^*_M) \text{ for } k = 1,2,3. \qquad (25)$$

On the other hand, using (6) for the Caputo derivative the following equalities can be obtained:

$$^{[0,t_M]}D^{\alpha}_{\,C}X_k(t_M) = \frac{1}{\Gamma(1-\alpha)} \cdot \int_0^{t_M} \frac{X_k{}'(\tau)}{(t_M-\tau)^{\alpha}} d\tau =$$

$$\frac{1}{\Gamma(1-\alpha)} \cdot \int_0^{t_{M_{O^*}}} \frac{X_k{}'(\tau)}{(t_M-\tau)^{\alpha}} d\tau + \frac{1}{\Gamma(1-\alpha)} \cdot \int_{t_{M_{O^*}}}^{t_M} \frac{X_k{}'(\tau)}{(t_M-\tau)^{\alpha}} d\tau =$$

$$\frac{1}{\Gamma(1-\alpha)} \cdot \int_0^{t_{M_{O^*}}} \frac{X_k{}'(\tau)}{(t_M-\tau)^{\alpha}} d\tau + \frac{1}{\Gamma(1-\alpha)} \cdot \int_0^{t_M-t_{M_{O^*}}} \frac{X_k{}'(t_{M_{O^*}}+\xi)}{(t_M-t_{M_{O^*}}-\xi)^{\alpha}} d\xi = \quad (26)$$

$$\frac{1}{\Gamma(1-\alpha)} \cdot \int_0^{t_{M_{O^*}}} \frac{X_k{}'(\tau)}{(t_M-\tau)^{\alpha}} d\tau + \frac{1}{\Gamma(1-\alpha)} \cdot \int_0^{t^*_M} \frac{X^*_k{}'(\xi)}{(t^*_M-\xi)^{\alpha}} d\xi =$$

$$\frac{1}{\Gamma(1-\alpha)} \cdot \int_0^{t_{M_{O^*}}} \frac{X_k{}'(\tau)}{(t_M-\tau)^{\alpha}} d\tau + {}^{[0,t^*_M]}D^{\alpha}_{\,C}X^*_k(t^*_M)$$

Formula (25) and (26) imply that Caputo velocity of order α is objective if and only if the following equalities hold:

$$\frac{1}{\Gamma(1-\alpha)} \cdot \int_0^{t_{M_{O^*}}} \frac{X_k{}'(\tau)}{(t_M-\tau)^{\alpha}} d\tau + {}^{[0,t^*_M]}D^{\alpha}_{\,C}X^*_k(t^*_M) = \sum_{i=1}^{i=3} a_{ik} {}^{[0,t^*_M]}D^{\alpha}_{\,C}X^*_i(t^*_M) \quad (27)$$

for $k=1,2,3$.

Remark now, that the above condition in general is not fulfilled. For instance: if $k=1$ and $a_{11}=1, a_{21}=0, a_{31}=0$ (see [5]), then (27) becomes:

$$\frac{1}{\Gamma(1-\alpha)} \cdot \int_0^{t_{M_{O^*}}} \frac{X_1{}'(\tau)}{(t_M-\tau)^{\alpha}} d\tau = 0 \quad (28)$$

Equality (28) in general is not true. So, Caputo velocity of order α is nonobjective.

Nonobjective Description of the Material Point Velocity Using Riemann-Liouville Fractional Order Derivative Defined with Formula (7)

For observer O the Riemann-Liouville velocity $^{R-L}\vec{V}^{\alpha}{}_{M}$ of order α at the moment of time M is the vector obtained translating in E_3 the vector

$$^{[0,t_M]}D^{\alpha}{}_{R-L}\vec{X}(t_M) = \sum_{i=1}^{i=3} {}^{[0,t_M]}D^{\alpha}{}_{R-L}X_i(t_M)\vec{e}_i \quad (29)$$

to the point of coordinates $(X_1(t_M), X_2(t_M), X_3(t_M))$. Here $^{[0,t_M]}D^{\alpha}{}_{R-L}X_i(t_M)$ is the Riemann-Liouville derivative of order α of the function $X_i(t_M)$ at t_M for $i = 1,2,3$ (formula (7)).

For observer O^* the Riemann-Liouville velocity $^{R-L}\vec{V}^{*\alpha}{}_{M}$ of order α at the moment of time M is the vector obtained translating in E_3 the vector:

$$^{[0,t^*_M]}D^{\alpha}{}_{R-L}\vec{X}^*(t^*_M) = \sum_{i=1}^{i=3} {}^{[0,t^*_M]}D^{\alpha}{}_{R-L}X^*_i(t^*_M)\vec{e}^*_i \quad (30)$$

at the point of coordinates $(X^*_1(t^*_M), X^*_2(t^*_M), X^*_3(t^*_M))$. Here $^{[0,t^*_M]}D^{\alpha}{}_{R-L}X^*_i(t^*_M)$ is the Riemann-Liouville derivative of order α of the function $X^*_i(t^*_M)$ at t^*_M for $i = 1,2,3$ (formula (7)).

The Riemann-Liouville velocity of order α is objective if and only if equality $^{R-L}\vec{V}^{\alpha}{}_{M} = {}^{R-L}\vec{V}^{*\alpha}{}_{M}$ holds.

Remark now that equality $^{R-L}\vec{V}^{\alpha}{}_{M} = {}^{R-L}\vec{V}^{*\alpha}{}_{M}$ holds if and only if:

$$\begin{aligned}{}^{[0,t_M]}D^{\alpha}{}_{R-L}\vec{X}(t_M) &= \sum_{i=1}^{i=3} {}^{[0,t_M]}D^{\alpha}{}_{R-L}X_i(t_M)\vec{e}_i = \\ {}^{[0,t^*_M]}D^{\alpha}{}_{R-L}\vec{X}^*(t^*_M) &= \sum_{i=1}^{i=3} {}^{[0,t^*_M]}D^{\alpha}{}_{R-L}X^*_i(t^*_M)\vec{e}^*_i\end{aligned} \quad (31)$$

Objective and Nonobjective Mathematical Description of ...   113

Hence, by using $\vec{e}^*{}_i = \sum_{k=1}^{k=3} a_{ik}\vec{e}_k$ the following equalities are obtained:

$$\sum_{i=1}^{i=3} {}^{[0,t_M]}D^\alpha_{R-L} X_i(t_M)\vec{e}_i = \sum_{i=1}^{i=3} {}^{[0,t^*_M]}D^\alpha_{R-L} X^*_i(t^*_M)\vec{e}^*{}_i =$$

$$= \sum_{i=1}^{i=3} {}^{[0,t^*_M]}D^\alpha_{R-L} X^*_i(t^*_M) \cdot \left(\sum_{k=1}^{k=3} a_{ik}\vec{e}_k\right) = \qquad (32)$$

$$= \sum_{k=1}^{k=3}\left(\sum_{i=1}^{i=3} a_{ik} \cdot {}^{[0,t^*_M]}D^\alpha_{R-L} X^*_i(t^*_M)\right)\vec{e}_k$$

So, the Riemann-Liouville velocity of order α, defined by (29), is objective if and only if the next equalities hold:

$$^{[0,t_M]}D^\alpha_{R-L} X_k(t_M) = \sum_{i=1}^{i=3} a_{ik} \cdot {}^{[0,t^*_M]}D^\alpha_{R-L} X^*_i(t^*_M) \qquad (33)$$

for $k = 1,2,3$.

On the other hand, using (7) for the Riemann-Liouville α order derivative the following equalities are obtained:

$$^{[0,t_M]}D^\alpha_{R-L} X_k(t_M) = \frac{1}{\Gamma(1-\alpha)} \cdot \frac{d}{dt_M}\int_0^{t_M} \frac{X_k(\tau)}{(t_M - \tau)^\alpha} d\tau =$$

$$= \frac{1}{\Gamma(1-\alpha)} \cdot \frac{d}{dt_M}\int_0^{t_{M_{O^*}}} \frac{X_k(\tau)}{(t_M - \tau)^\alpha} d\tau +$$

$$\frac{1}{\Gamma(1-\alpha)} \cdot \frac{d}{dt_M}\int_{t_{M_{O^*}}}^{t_M} \frac{X_k(\tau)}{(t_M - \tau)^\alpha} d\tau =$$

$$= \frac{1}{\Gamma(1-\alpha)} \cdot \frac{d}{dt_M}\int_0^{t_{M_{O^*}}} \frac{X_k(\tau)}{(t_M - \tau)^\alpha} d\tau +$$

$$\frac{1}{\Gamma(1-\alpha)} \cdot \frac{d}{dt_M}\int_0^{t_M - t_{M_{O^*}}} \frac{X_k(t_{M_{O^*}} + \xi)}{(t_M - t_{M_{O^*}} - \xi)^\alpha} d\xi =$$

$$= \frac{1}{\Gamma(1-\alpha)} \cdot \frac{d}{dt_M}\int_0^{t_{M_{O^*}}} \frac{X_k(\tau)}{(t_M - \tau)^\alpha} d\tau +$$

$$\frac{1}{\Gamma(1-\alpha)} \cdot \frac{d}{dt^*_M}\int_0^{t^*_M} \frac{X^*_k(\xi)}{(t^*_M - \xi)^\alpha} d\xi = \qquad (34)$$

$$= \frac{1}{\Gamma(1-\alpha)} \cdot \frac{d}{dt_M}\int_0^{t_{M_{O^*}}} \frac{X_k(\tau)}{(t_M - \tau)^\alpha} d\tau + {}^{[0,t^*_M]}D^\alpha_{R-L} X^*_k(t^*_M)$$

Formula (33) and (34) imply that the Riemann-Liouville velocity of order α is objective if and only if the following equalities hold:

$$\frac{1}{\Gamma(1-\alpha)} \cdot \frac{d}{dt_M} \int_0^{t_{M_{O^*}}} \frac{X_k(\tau)}{(t_M-\tau)^\alpha} d\tau + {}^{[0,t_{M^*}]}D^\alpha_{R-L} X^*_k(t^*_M) = \sum_{i=1}^{i=3} a_{ik} \cdot {}^{[0,t_{M^*}]}D^\alpha_{R-L} X^*_i(t^*_M) \quad (35)$$

for $k = 1,2,3$.

The above conditions in general are not fulfilled. For instance if $k=1$ and $a_{11}=1, a_{21}=0, a_{31}=0$ (see [5]), then (35) becomes

$$\frac{1}{\Gamma(1-\alpha)} \cdot \frac{d}{dt_M} \int_0^{t_{M_{O^*}}} \frac{X_1(\tau)}{(t_M-\tau)^\alpha} d\tau = 0 \quad (36)$$

This equality is not true in general. So, Riemann-Liouville velocity of order α is not objective.

Objective Description of the Material Point Acceleration with Second Order Derivatives

For observer O the acceleration $\vec{A}_M$ at the moment of time M is the vector obtained translating in E_3 the vector $\vec{X}''(t_M) = \sum_{i=1}^{i=3} X''_i(t_M) \vec{e}_i$ to the point of coordinates $(X_1(t_M), X_2(t_M), X_3(t_M))$. Here $X''_i(t_M)$ is the second order derivative of the function $X_i(t_M)$ at t_M for $i=1,2,3$.

Observer O^* describes the acceleration $\vec{A}^*_M$ at the moment of time M similarly.

The acceleration description is objective if and only if $\vec{A}_M = \vec{A}*_M$.
Equality $\vec{A}_M = \vec{A}*_M$ can be proven differentiating twice in (3) and obtaining:

$$X''_k(t_M) = \sum_{i=1}^{i=3} a_{ik} \cdot X*''_i(t*_M) \text{ for } k = 1,2,3 \tag{37}$$

Equalities (37) show that equality $\vec{A}_M = \vec{A}*_M$ holds. So, the acceleration description with second order derivatives is objective.

Objective Description of the Material Point Acceleration Using General Liouville-Caputo Fractional Order Derivative Defined with Formula (8)

Observer O describes the general Liouville-Caputo acceleration of order $\beta = 1 + \alpha$, $^{L-C}\vec{A}^\beta_M$ at the moment of time M translating in E_3 the vector $^{(-\infty,t_M]}D^\beta_{L-C}\vec{X}(t_M) = \sum_{i=1}^{i=3} {}^{(-\infty,t_M]}D^\beta_{L-C}X_i(t_M)\vec{e}_i$ in the point of coordinates $(X_1(t_M), X_2(t_M), X_3(t_M))$. Here $^{(-\infty,t_M]}D^\beta_{L-C}X_i(t_M)$ is the general Liouville-Caputo derivative of order $\beta = 1 + \alpha$ of the function $X_k(t_M)$ at t_M for $k = 1,2,3$ and is given by:

$$^{(-\infty,t_M]}D^\beta_{L-C}X_k(t_M) = \frac{1}{\Gamma(2-\alpha)} \cdot \int_{-\infty}^{t_M} \frac{X_k''(\tau)}{(t_M - \tau)^{\beta-1}} d\tau \tag{38}$$

Observer $O*$ describes the general Liouville-Caputo acceleration of order $\beta = 1 + \alpha$, $^{L-C}\vec{A}^\beta *_M$ at the moment of time M similarly.

The general Liouville-Caputo acceleration of order $\beta = 1 + \alpha$ is independent on observer if and only if $^{L-C}\vec{A}^\beta_M = {}^{L-C}\vec{A}^\beta *_M$. The proof of this equality is similar with the proof of the objectivity of general Liouville-Caputo fractional order velocity. So, the general Liouville- Caputo acceleration is objective.

Objective Description of the Material Point Acceleration Using General Riemann-Liouville Fractional Order Derivative Defined with Formula (9)

Observer O describes the general Riemann-Liouville acceleration of order $\beta = 1+\alpha$, $^{R-L}\vec{A}^{\beta}{}_{M}$ at the moment of time M translating in E_3 the vector $^{(-\infty,t_M]}D_{R-L}^{\beta}\vec{X}(t_M) = \sum_{i=1}^{i=3}{}^{(-\infty,t_M]}D_{R-L}^{\beta}X_i(t_M)\vec{e}_i$ in the point of coordinates $(X_1(t_M), X_2(t_M), X_3(t_M))$. Here $^{(-\infty,t_M]}D_{R-L}^{\beta}X_k(t_M)$ is the general Riemann-Liouville, $\beta = 1+\alpha$ order derivative of the function $X_k(t_M)$ at t_M for $k = 1,2,3$ and is given by:

$$^{(-\infty,t_M]}D_{R-L}^{\beta}X_k(t_M) = \frac{1}{\Gamma(2-\alpha)} \cdot \frac{d^2}{dt_M^2} \int_{-\infty}^{t_M} \frac{X_k(\tau)}{(t_M-\tau)^{\beta-1}} d\tau \qquad (39)$$

Using observer O^* description, the general Riemann-Liouville acceleration of order $\beta = 1+\alpha$, $^{R-L}\vec{A}^{\beta}*_M$ at the moment of time M of the material point is defined similarly.

The general Riemann-Liouville acceleration of order $\beta = 1+\alpha$ is independent on observer, if and only if $^{R-L}\vec{A}^{\beta}{}_M = {}^{R-L}\vec{A}^{\beta}*_M$. The proof of this equality is similar with the proof of the objectivity of general Liouville-Caputo fractional order velocity. So, the general Riemann-Liouville acceleration is objective.

Nonobjective Description of the Material Point Acceleration Using Caputo Fractional Order Derivative Defined with Formula (6)

Observer O describes the Caputo acceleration of order $\beta = 1+\alpha$, $^C\vec{A}^\beta{}_M$ at the moment of time M translating in E_3 the vector $^{[0,t_M]}D_C^\beta \vec{X}(t_M) = \sum_{i=1}^{i=3} {}^{[0,t_M]}D_C^\beta X_i(t_M)\vec{e}_i$ to the point of coordinates $(X_1(t_M), X_2(t_M), X_3(t_M))$. Here $^{[0,t_M]}D_C^\beta X_i(t_M)$ is the Caputo derivative of order $\beta = 1+\alpha$ of the function $X_k(t_M)$ at t_M for $k = 1,2,3$ and is given by:

$$^{[0,t_M]}D_{C-L}^\beta X_k(t_M) = \frac{1}{\Gamma(2-\alpha)} \cdot \int_0^{t_M} \frac{X_k''(\tau)}{(t_M-\tau)^{\beta-1}} d\tau \qquad (40)$$

Observer O^* describes the Caputo acceleration of order $\beta = 1+\alpha$, $^C\vec{A}^\beta *_M$ at the moment of time M similarly.

The Caputo acceleration of order $\beta = 1+\alpha$ is objective if and only if $^C\vec{A}^\beta{}_M = {}^C\vec{A}^\beta *_M$. In general, this equality is not valid. This can be proven analogously with the nonobjectivity of the Caputo velocity of order α. So, the Caputo fractional order acceleration is nonobjective.

Nonobjective Description of the Material Point Acceleration Using Riemann-Liouville Fractional Order Derivative Defined with Formula (7)

Observer O describes the Riemann-Liouville acceleration of order $\beta = 1+\alpha$, $^{R-L}\vec{A}^\beta{}_M$ at the moment of time M translating in E_3 the vector $^{[0,t_M]}D_{R-L}^\beta \vec{X}(t_M) = \sum_{i=1}^{i=3} {}^{[0,t_M]}D_{R-L}^\beta X_i(t_M)\vec{e}_i$ to the point of coordinates

$(X_1(t_M), X_2(t_M), X_3(t_M))$. Here ${}^{[0,t_M]}D_{R-L}^{\beta}X_k(t_M)$ is the Riemann-Liouville, $\beta = 1+\alpha$ order derivative of the function $X_k(t_M)$ at t_M for $k=1,2,3$ and is given by:

$${}^{[0,t_M]}D_{R-L}^{\beta}X_k(t_M) = \frac{1}{\Gamma(2-\alpha)} \cdot \frac{d^2}{dt_M^2}\int_0^{t_M}\frac{X_k(\tau)}{(t_M-\tau)^{\beta-1}}d\tau \qquad (41)$$

Observer O^* describes the Riemann-Liouville acceleration of order $\beta = 1+\alpha$, ${}^{R-L}\vec{A}^{\beta}*_M$ at the moment of time M similarly.

The Riemann-Liouville acceleration of order $\beta = 1+\alpha$ is objective if and only if ${}^{R-L}\vec{A}^{\beta}{}_M = {}^{R-L}\vec{A}^{\beta}*_M$. In general, this equality is not valid. This can be proven analogously with the nonobjectivity of the Caputo acceleration of order α. So, the Riemann-Liouville fractional order acceleration is nonobjective.

Objective Description of the Material Point Dynamics Using First and Second Order Derivative

The verbal expression of the second law of Newton is "The rate of change of momentum is proportional to the impressed force, and takes place in the direction of the straight line in which the force acts" [2]. In terms of the observer O description this means that the function $\vec{X}(t_M) = \sum_{i=1}^{i=3}X_i(t_M)\vec{e}_i$, describing the motion with respect to R_O, verifies the following differential equation:

$$m \cdot \vec{X}''(t_M) = \vec{F}_O(t_M, \vec{X}(t_M), \vec{X}'(t_M)) \qquad (42)$$

Here: m represents the material point mass, t_M represents the moment of time, $\vec{X}(t_M)$ represents the position of the material point with respect to R_O,

$\vec{X}'(t_M)$ represents the material point velocity with respect to R_O, $\vec{X}''(t_M)$ represents the material point acceleration with respect to R_O, and $\vec{F}_O$ is the vector valued function depending on the moment of time, position and velocity, representing with respect to R_O the force field $\vec{F}$ acting on the material point.

In terms of the observer O^* description, the same verbal expression leads to the conclusion that the function $\vec{X}^*(t^*_M) = \sum_{i=1}^{i=3} X^*_i(t^*_M)\vec{e}^*_i$ describing the motion with respect to R_{O^*}, under the action of the same force field, verifies the following differential equation:

$$m \cdot \vec{X}^{*''}(t^*_M) = \vec{F}_{O^*}(t^*_M, \vec{X}^*(t^*_M), \vec{X}^{*'}(t_M)) \qquad (43)$$

Here: m represents the material point mass, t^*_M represents the moment of time, $\vec{X}^*(t^*_M) = \sum_{i=1}^{i=3} X^*_i(t^*_M)\vec{e}^*_i$ represents the position of the material point with respect to R_{O^*}, $\vec{X}^{*'}(t^*_M)$ represents the velocity of the material point with respect to R_{O^*}, $\vec{X}^{*''}(t^*_M)$ represents the acceleration of the material point with respect to R_{O^*}, and $\vec{F}_{O^*}$ is the vector valued function depending on the moment of time, position, and velocity, which represents with respect to R_{O^*} the same force field $\vec{F}$, acting on the material point.

Because the functions $\vec{F}_O$ and $\vec{F}_{O^*}$ represent the same force field $\vec{F}$ in two different reference frames the components of the vector valued functions $\vec{F}_O$ and $\vec{F}_{O^*}$ verify the following relations:

$$F_{O^*k} = a_{k1}F_{O1} + a_{k2}F_{O2} + a_{k3}F_{O3}, \qquad k=1,2,3 \qquad (44)$$

Objectivity of the dynamics description means that the solutions of the differential equations (42) and (43) describe the same movement of the material point under the action of the force field $\vec{F}$. This can be proven showing that if $\vec{X}(t_M) = \sum_{i=1}^{i=3} X_i(t_M)\vec{e}_i$ is a solution of (42), then the function $\vec{X}*(t*_M) = \vec{X}*_O + \vec{X}(t_M)$, is a solution of (43) and vice versa, if $\vec{X}*(t*_M) = \sum_{i=1}^{i=3} X*_i(t*_M)\vec{e}*_i$ is a solution of (43), then the function $\vec{X}(t_M) = \vec{X}_{O*} + \vec{X}*(t*_M)$, is a solution of (42).

In other words, the dynamics of the material point can be described by the system (42) or by the system (43). This means, that the dynamics description is independent on the observer, so it is objective.

Objective Description of the Material Point Dynamics Using General Liouville-Caputo Fractional Order Derivative Defined with Formula (8)

In terms of the observer O and general Liouville-Caputo fractional order derivative the verbal expression of the second law of Newton means that the function $\vec{X}(t_M) = \sum_{i=1}^{i=3} X_i(t_M)\vec{e}_i$, describing the motion with respect to R_O, under the action of the force field $\vec{F}$, verifies the following differential equation:

$$m \cdot {}^{(-\infty, t_M]}D^{\beta}_{L-C}\vec{X}(t_M) = \vec{F}_O(t_M, \vec{X}(t_M), {}^{(-\infty, t_M]}D^{\alpha}_{L-C}\vec{X}(t_M)) \qquad (45)$$

Here: $\beta = 1 + \alpha$, $0 < \alpha < 1$, m represents the material point mass, t_M represents the moment of time, $\vec{X}(t_M)$ represents the position of the material point with respect to R_O, ${}^{(-\infty, t_M]}D^{\alpha}_{C-L}\vec{X}(t_M)$, represents the general Liouville-

Caputo velocity of order α with respect to R_O, $^{(-\infty,t_M^*]}D^\beta_{L-C}\vec{X}(t_M)$ represents the general Liouville-Caputo acceleration of order β with respect to R_O, and $\vec{F}_O$ is the vector valued function depending on the moment of time, position and general Liouville-Caputo velocity, representing with respect to R_O the force field $\vec{F}$, acting on the material point.

In terms of the observer O^* and general Liouville-Caputo fractional order derivative, the same verbal expression means that the function $\vec{X}*(t*_M) = \sum_{i=1}^{i=3} X*_i(t*_M)\vec{e}*$, describing the motion with respect to R_O*, under the action of the same force field $\vec{F}$, verifies the following differential equation:

$$m \cdot {}^{(-\infty,t^*_M]}D^\beta_{L-C}\vec{X}*(t*_M) = \vec{F}_{O*}(t*_M, \vec{X}*(t*_M), {}^{(-\infty,t^*_M]}D^\alpha_{L-C}\vec{X}*(t*_M)) \qquad (46)$$

Here: $\beta = 1+\alpha$, $0 < \alpha < 1$, m represents the material point mass, $t*_M$ represents the moment of time, $\vec{X}*(t*_M)$ represents the position of the material point with respect to R_{O*}, $^{(-\infty,t^*_M]}D^\alpha_{L-C}\vec{X}*(t*_M)$ represents the general Liouville-Caputo velocity of order α with respect to R_{O*}, $^{(-\infty,t^*_M]}D^\beta_{L-C}\vec{X}*(t*_M)$ represents the acceleration of order β with respect to R_{O*}, and $\vec{F}_{O*}$ is the vector valued function, depending on the moment of time, position and general Liouville-Caputo velocity of order α, representing with respect to R_{O*} the same force field $\vec{F}$ acting on the material point.

Because the functions $\vec{F}_O$ and $\vec{F}_{O*}$ represent the same force field $\vec{F}$ in two different reference frames the components of the vector valued functions $\vec{F}_O$ and $\vec{F}_{O*}$ verify the following relations:

$$F_{O*k} = a_{k1}F_{O1} + a_{k2}F_{O2} + a_{k3}F_{O3}, \qquad k = 1, 2, 3 \qquad (47)$$

The dynamics description is objective if and only if the differential equations (44) and (45) describe the same movement under the action of the force field $\vec{F}$. This can be proven showing that: if $\vec{X}(t_M) = \sum_{i=1}^{i=3} X_i(t_M)\vec{e}_i$ is a solution of (45), then the function $\vec{X}*(t*_M) = \vec{X}*_O + \vec{X}(t_M)$, is a solution of (46) and vice versa, if $\vec{X}*(t*_M) = \sum_{i=1}^{i=3} X*_i(t*_M)\vec{e}*_i$ is a solution of (46), then the function $\vec{X}(t_M) = \vec{X}_{O*} + \vec{X}*(t*_M)$ is a solution of (45).

In other words, the dynamics of the material point can be described by the system (45) or by the system (46). This means, that the dynamics description is independent on the observer, so it is objective.

Objective Description of the Material Point Dynamics Using General Riemann-Liouville Fractional Order Derivative Defined with Formula (9)

In terms of the observer O and general Riemann-Liouville fractional order derivative the verbal expression of the second law of Newton means that the function $\vec{X}(t_M) = \sum_{i=1}^{i=3} X_i(t_M)\vec{e}_i$, describing the motion with respect to R_O, under the action of the force field $\vec{F}$, verifies the following differential equation:

$$m \cdot {}^{(-\infty, t_M]}D^{\beta}_{R-L}\vec{X}(t_M) = \vec{F}_o(t_M, \vec{X}(t_M), {}^{(-\infty, t_M]}D^{\alpha}_{R-L}\vec{X}(t_M), {}^{(-\infty, t_M]}) \qquad (48)$$

Here: $\beta = 1+\alpha$, $0 < \alpha < 1$, m represents the mass of material point, t_M represents the moment of time, $\vec{X}(t_M)$ represents the position of the material point with respect to R_O, $^{(-\infty, t_M]}D^\alpha_{R-L}\vec{X}(t_M)$ represents the general Riemann-Liouville velocity of order α with respect to R_O, $^{(-\infty, t_M]}D^\beta_{R-L}\vec{X}(t_M)$ represents the general Riemann-Liouville acceleration of order β with respect to R_O, and $\vec{F}_O$ is the vector valued function depending on the moment of time, position and general Riemann-Liouville velocity, representing with respect to R_O the force field $\vec{F}$, acting on the material point.

In terms of the observer $O*$ and general Riemann-Liouville fractional order derivative, the same verbal expression means that the function $\vec{X}*(t*_M) = \sum_{i=1}^{i=3} X*_i(t*_M)\vec{e}*_i$, describing the motion with respect to R_O under the action of the same force field $\vec{F}$, verifies the following differential equation:

$$m \cdot {}^{(-\infty, t*_M]}D^\beta_{R-L}\vec{X}(t*_M) = \vec{F}_{O*}(t*_M, \vec{X}(t*_M), {}^{(-\infty, t*_M]}D^\alpha_{R-L}\vec{X}(t*_M)) \qquad (49)$$

Here: $\beta = 1+\alpha$, $0 < \alpha < 1$, m represents the mass of material point, $t*_M$ represents the moment of time, $\vec{X}*(t*_M)$ represents the position of the material point with respect to R_{O*}, $^{(-\infty, t*_M]}D^\alpha_{R-L}\vec{X}*(t*_M)$ represents the general Riemann-Liouville velocity of order α with respect to R_{O*}, $^{(-\infty, t*_M]}D^\beta_{R-L}\vec{X}*(t*_M)$ represents the general Riemann-Liouville acceleration of order β with respect to R_{O*}, and $\vec{F}_{O*}$ is the vector valued function, depending on the moment of time, position, and general Riemann-Liouville velocity of order α, representing with respect to R_{O*} the same force field $\vec{F}$ acting on the material point.

Because the functions $\vec{F}_O$ and $\vec{F}_{O*}$ represent the same force field $\vec{F}$ in two different reference frames, the components of the vector valued functions $\vec{F}_O$ and $\vec{F}_{O*}$ verify the following relations:

$$F_{O*k} = a_{k1}F_{O1} + a_{k2}F_{O2} + a_{k3}F_{O3}, \qquad k = 1,2,3 \qquad (50)$$

The dynamics description is objective if and only if differential equations (48) and (49) describe the same movement under the action of the force field $\vec{F}$. This can be proven showing that if $\vec{X}(t_M) = \sum_{i=1}^{i=3} X_i(t_M)\vec{e}_i$ is a solution of (48), then the function $\vec{X}*(t*_M) = \vec{X}*_O + \vec{X}(t_M)$ is a solution of (49) and vice versa, if $\vec{X}*(t*_M) = \sum_{i=1}^{i=3} X*_i(t*_M)\vec{e}*_i$ is a solution of (49), then the function $\vec{X}(t_M) = \vec{X}_{O*} + \vec{X}*(t*_M)$ is a solution of (48).

In other words, the dynamics of the material point can be described by the system (48) or by the system (49). This means, that the dynamics description is independent on the observer, so it is objective.

Nonobjective Description of the Material Point Dynamics Using Caputo Fractional Order Derivative Defined with Formula (6)

In terms of the description of observer O and Caputo fractional order derivative, the second law of Newton means that the function $\vec{X}(t_M) = \sum_{i=1}^{i=3} X_i(t_M)\vec{e}_i$ describing the motion with respect to R_O, under the action of a force field $\vec{F}$, verifies the following differential equation:

$$m \cdot {}^{[0,t_M]}D^\beta_C \vec{X}(t_M) = \vec{F}_O(t_M, \vec{X}(t_M), {}^{[0,t_M]}D^\alpha_C \vec{X}(t_M)) \qquad (51)$$

Here: $\beta = 1+\alpha$, $0 < \alpha < 1$, m represents the mass of material point, t_M represents the moment of time, $\vec{X}(t_M)$ represents the position of the material point with respect to R_O, $^{[0,t_M]}D^\alpha{}_c\vec{X}(t_M)$ represents the fractional Caputo velocity of order α with respect to R_O, $^{[0,t_M]}D^\beta{}_c\vec{X}(t_M)$ represents the fractional Caputo acceleration of order β with respect to R_O and $\vec{F}_O$ is the vector valued function, depending on the moment of time, position, and Caputo velocity, representing with respect to R_O the force field $\vec{F}$, acting on the material point.

In terms of the observer O^* and Caputo fractional order derivative, the same verbal expression means that the function

$$\vec{X}*(t*_M) = \sum_{i=1}^{i=3} X*_i(t*_M)\vec{e}*$$

describing the motion with respect to R_{O^*}, under the action of the same force field $\vec{F}$, verifies the following differential equation:

$$m \cdot {}^{[0,t*_M]}D^\beta{}_c\vec{X}*_k(t*_M) = \vec{F}_{O^*}(t*_M, \vec{X}*(t*_M), {}^{[0,t*_M]}D^\alpha{}_c\vec{X}*(t*_M)) \quad (52)$$

Here: $\beta = 1+\alpha$, $0 < \alpha < 1$, m represents the mass of material point, $t*_M$ represents the moment of time, $\vec{X}*(t*_M)$ represents the position of the material point with respect to R_{O^*}, $^{[0,t*_M]}D^\alpha{}_c\vec{X}*(t*_M)$ represents the Caputo velocity of order α with respect to R_{O^*}, $^{[0,t*_M]}D^\beta{}_c\vec{X}*(t*_M)$ represents the Caputo acceleration of order β with respect to R_{O^*}, and $\vec{F}_{O^*}$ is the vector valued function, depending on the moment of time, position, and Caputo velocity of order α, representing with respect to R_{O^*} the same force field $\vec{F}$ acting on the material point.

Because the functions $\vec{F}_O$ and $\vec{F}_{O^*}$ represent the same force field $\vec{F}$ in two different reference frames the components of the vector valued functions $\vec{F}_O$ and $\vec{F}_{O^*}$ verify the following relations:

$$F_{O^*k} = a_{k1}F_{O1} + a_{k2}F_{O2} + a_{k3}F_{O3}, \qquad k = 1,2,3 \qquad (53)$$

The dynamics description is objective if and only if differential equations (51) and (52) describe the same movement under the action of the force field $\vec{F}$. The statement that equations (51) and (52) describe the same movement under the action of the force field $\vec{F}$ in general is not valid. This is shown in [5] in the case of a force field which acts along a line L and depends only on the position. So, the dynamic description in this case is nonobjective.

Nonobjective Description of the Material Point Dynamics Using Riemann-Liouville Fractional Order Derivative Defined with Formula (7)

In terms of the description of observer O and Riemann-Liouville fractional order derivative, the second law of Newton means that the function $\vec{X}(t_M) = \sum_{i=1}^{i=3} X_i(t_M)\vec{e}_i$ describing the motion with respect to R_O under the action of a force field $\vec{F}$, verifies the following differential equation:

$$m \cdot {}^{[0,t_M]}D^{\beta}_{R-L}\vec{X}(t_M) = \vec{F}_O(t_M, \vec{X}(t_M), {}^{[0,t_M]}D^{\alpha}_{R-L}\vec{X}(t_M)) \qquad (54)$$

Here: $\beta = 1 + \alpha$, $0 < \alpha < 1$, m represents the mass of material point, t_M represents the moment of time, $\vec{X}(t_M)$ represents the position of the material point with respect to R_O, ${}^{[0,t_M]}D^{\alpha}_{R-L}\vec{X}(t_M)$ represents the fractional

Riemann-Liouville velocity of order α with respect to R_O, ${}^{[0,t_M]}D^{\beta}_{R-L}\vec{X}(t_M)$ represents the fractional Riemann-Liouville acceleration of order β with respect to R_O and $\vec{F}_O$ is the vector valued function, depending on the moment of time position and Riemann-Liouville velocity, representing with respect to R_O the force field $\vec{F}$, acting on the material point.

In terms of the observer O^* and Riemann-Liouville fractional order derivative, the same verbal expression means that the function $\vec{X}*(t*_M) = \sum_{i=1}^{i=3} X*_i(t*_M)\vec{e}*$, describing the motion with respect to R_{O^*} under the action of the same force field $\vec{F}$, verifies the following differential equation:

$$m.{}^{[0,t^*_M]}D^{\beta}_{R-L}\vec{X}*(t*_M) = \vec{F}_{O^*}(t*_M, \vec{X}*(t*_M), {}^{[0,t^*_M]}D^{\alpha}_{R-L}\vec{X}*(t*_M)) \qquad (55)$$

Here: $\beta = 1+\alpha$, $0 < \alpha < 1$, m represents the mass of material point, $t*_M$ represents the moment of time, $\vec{X}*(t*_M)$ represents the position of the material point with respect to R_{O^*}, ${}^{[0,t^*_M]}D^{\alpha}_{R-L}\vec{X}*(t*_M)$ represents the Riemann-Liouville velocity of order α with respect to R_{O^*}, ${}^{[0],t^*_M]}D^{\beta}_{R-L}\vec{X}*(t*_M)$ represents the Riemann-Liouville acceleration of order β with respect to R_{O^*}, and $\vec{F}_{O^*}$ is the vector valued function, depending on the moment of time, position, and Riemann-Liouville velocity of order α, representing with respect to R_{O^*} the same force field $\vec{F}$ acting on the material point.

Because the functions $\vec{F}_O$ and $\vec{F}_{O^*}$ represent the same force field $\vec{F}$ in two different reference frames the components of the vector valued functions $\vec{F}_O$ and $\vec{F}_{O^*}$ verify the following relations:

$$F_{O^*k} = a_{k1}F_{O1} + a_{k2}F_{O2} + a_{k3}F_{O3}, \qquad k = 1,2,3 \tag{56}$$

The dynamics description is objective if and only if differential equations (54) and (55) describes the same movement under the action of the force field $\vec{F}$. The statement that equations (54) and (55) describes the same movement under the action of the force field $\vec{F}$ in general is not valid. This is shown in [5] in the case of a force field which acts along a line L and depends only on the position. So the dynamics description in this case is nonobjective.

CONCLUSION

1. Velocity description with fractional order Caputo derivative (6) is nonobjective, but velocity description with fractional order general Liouville-Caputo derivative (8) or first order derivative is objective.
2. Velocity description with Riemann-Liouville fractional order derivative (7) is nonobjective, but velocity description with fractional order general Riemann-Liouville derivative (9) is objective.
3. Acceleration description with fractional order Caputo derivative (6) is nonobjective, but acceleration description with fractional order general Liouville-Caputo derivative (8) or second order derivative is objective.
4. Acceleration description with Riemann-Liouville fractional order derivative (7) is nonobjective, but acceleration description with fractional order general Riemann-Liouville derivative (9) is objective.
5. Dynamics description with fractional order Caputo derivative (6) is nonobjective, but dynamics description with fractional order general Liouville-Caputo derivative (8) or first and second order derivative is objective.
6. Dynamics description with Riemann-Liouville fractional order derivative (7) is nonobjective, but dynamics description with fractional order general Riemann-Liouville derivative (9) is objective.

CONFLICT OF INTEREST

This research did not receive any specific grant from funding agencies in the public, commercial or not-for-profit sectors.

REFERENCES

[1] Debs, Talal A., and Redhead, Michael L. G. 2007. *Objectivity, Invariance, and Convention: Symmetry in Physical Science*. Cambridge: Harvard University Press.

[2] Chester, W. 1991. *Mechanics*. London: Chapman & Hall.

[3] Demirci, Elif and Ozalp, Nuri 2012. "A method for solving differential equations of fractional order." *Journal of Computational and Applied Mathematics*, 236: 2754-2762.

[4] Ortigueira, Manuel D., and Machado, Jose A. Tenreiro. 2017. Which Derivative?" *Fractal and Fractional*. 2017, 1, 3; doi:10.3390/fractalfract1010003.

[5] Balint, Agneta M. and Balint, Stefan. 2019. "Description of Mechanical Movements Using Fractional Order Derivatives is not Objective." *INCAS Bulletin*, 11(2):15-28.

INDEX

A

axons, viii, 55, 77, 78

B

batteries, 69

C

Caputo, viii, 55, 61, 62, 63, 67, 71, 72, 73, 75, 80, 81, 82, 86, 87, 90, 92, 93, 94, 95, 96, 97, 99, 103, 104, 105, 109, 110, 111, 115, 116, 117, 118, 121, 124, 125, 128
Caputo fractional order, 61, 63, 67, 71, 72, 75, 80, 81, 86, 87, 90, 92, 109, 117, 124, 125
Chaos and Dissipation, 25
classical limit, 2, 3, 7, 22, 24, 25, 26, 27, 30, 32, 33, 36, 37, 53
classical mechanics, 100
classical-quantum transition, 2, 11, 24, 26, 52
conductance, 74, 91
conductivity, 95

D

dendrites, 77, 78
density matrix, vii, viii, 1, 25, 26, 31, 35, 50
derivatives, vii, viii, 55, 62, 63, 64, 65, 71, 72, 75, 76, 77, 80, 81, 82, 84, 86, 87, 88, 94, 95, 105, 115
dielectrics, 95
differential equations, 69, 70, 74, 84, 85, 120, 124, 126, 128
dynamical systems, 95

E

electric charge, vii, 55, 56
electron, 56, 58, 60, 62, 63, 65, 94
equality, 67, 68, 69, 72, 73, 74, 81, 82, 83, 85, 87, 88, 89, 90, 91, 92, 93, 105, 106, 108, 110, 112, 114, 115, 116, 117, 118
evolution, vii, viii, 79, 101

F

force, 118, 119, 120, 121, 122, 123, 124, 125, 126, 127, 128
fractional order derivative, vii, viii, 55, 56, 61, 62, 64, 67, 68, 71, 72, 73, 75, 76, 77, 80, 86, 90, 92, 93, 100, 103, 105, 107, 112, 115, 116, 117, 120, 122, 125, 126, 129

G

Galileo, 100

H

Hamiltonian, vii, viii, 1, 2, 3, 4, 5, 6, 7, 19, 21, 23, 25, 26, 27, 28, 29, 31, 32, 38, 51, 52

Index

I

Information Theory (IT), 3, 27
ion transport, 65, 67, 68, 69, 77, 78, 84, 92, 94, 95
ions, 69, 70, 84

J

Jaynes-Cummings semi-classical model, 2, 26

K

Kirchhoff voltage law, 62

L

Liouville-Caputo fractional order, viii, 55, 64, 68, 73, 76, 82, 88, 92, 94, 95, 104, 105, 106, 107, 115, 116, 120, 121, 128

M

mass, 118, 119, 120, 121, 122, 123, 125, 126, 127
material point, viii, 99, 100, 101, 102, 104, 105, 107, 109, 112, 114, 115, 116, 117, 118, 119, 120, 121, 122, 123, 124, 125, 126, 127
MaxEnt Density Matrix, vii, viii, 25, 26, 30
measurement, viii, 56, 78, 101

N

neurons, 90, 91

O

objectivity of a mathematical description, 56, 100
oscillation, 75, 76, 77

P

Principle of Uncertainty, viii, 25

Q

quantum dissipation, 2, 22, 23, 26, 29, 30, 51
quantum measurement, 2, 21, 26
quantum-classical transition zone, 9, 10, 36
quasi-periodicity, 9, 10, 36

R

reference frame, 78, 101, 103, 119, 121, 124, 126, 127
Riemann-Liouville fractional order, viii, 55, 62, 64, 65, 67, 68, 71, 72, 75, 76, 77, 80, 81, 82, 83, 86, 87, 88, 89, 92, 93, 94, 95, 99, 100, 103, 104, 107, 108, 109, 112, 113, 114, 116, 117, 118, 122, 123, 126, 127, 128
RLC, 56, 57, 58, 59, 60, 62
RLC series circuit, 56, 57, 58, 59

S

semiclassical, 2, 8, 23, 25, 26, 33, 36, 51
semiquantum, 2, 3, 5, 7, 9, 11, 13, 15, 17, 19, 21, 22, 23, 25, 26, 27, 29, 31, 33, 35, 37, 38, 39, 41, 43, 45, 47, 49, 50, 51, 53

T

transport, vii, viii, 55, 56, 57, 58, 60, 62, 63, 65, 69, 70, 71, 72, 73, 74, 80, 81, 82, 83, 84, 86, 87, 88, 89, 90, 91, 92, 93, 94, 95

V

vector, 104, 105, 106, 107, 109, 112, 114, 115, 116, 117, 119, 121, 123, 124, 125, 126, 127
velocity, 104, 105, 106, 107, 108, 109, 110, 111, 112, 113, 114, 116, 117, 119, 121, 123, 125, 127, 128